The Goddess – La Déesse

Investigations on the Legendary Citroën DS

Christian Sumi

The Goddess – La Déesse
Investigations on the Legendary Citroën DS

Lars Müller Publishers

Introduction

Introduction

· Traction Avant (1934)
· 2CV (1949)
· DS (1955)
· Ami 6 (1961)

Roland Barthes could not resist the fascination of the Citroën DS, "la Déesse" (the Goddess). His terse essay in *Mythologies* (1957; English translation 1979) has influenced the car's reception over the course of years. The sentence most often cited reads, *The DS – the Goddess – has all the features of one of those objects from another universe.*[1] The car became the ultimate icon of the French lifestyle and was extremely popular. Ironically, it was designed by an Italian from Varese, Flaminio Bertoni, "Créator de formes" (creator of forms).[2] He designed four cars for Citroën, all of which became icons of automobile design: Traction Avant (1934), 2CV (1949), Citroën DS (1955) and Ami 6 (1961).

1 – This and the following quotations are cited from: Roland Barthes, *The Eiffel Tower and Other Mythologies,* translated by Richard Howard, New York, Hill and Wang, 1979.
2 – This is the title of a typed manuscript in the Flaminio Bertoni Archive. For more on Bertoni, see the appendix: "Actors."

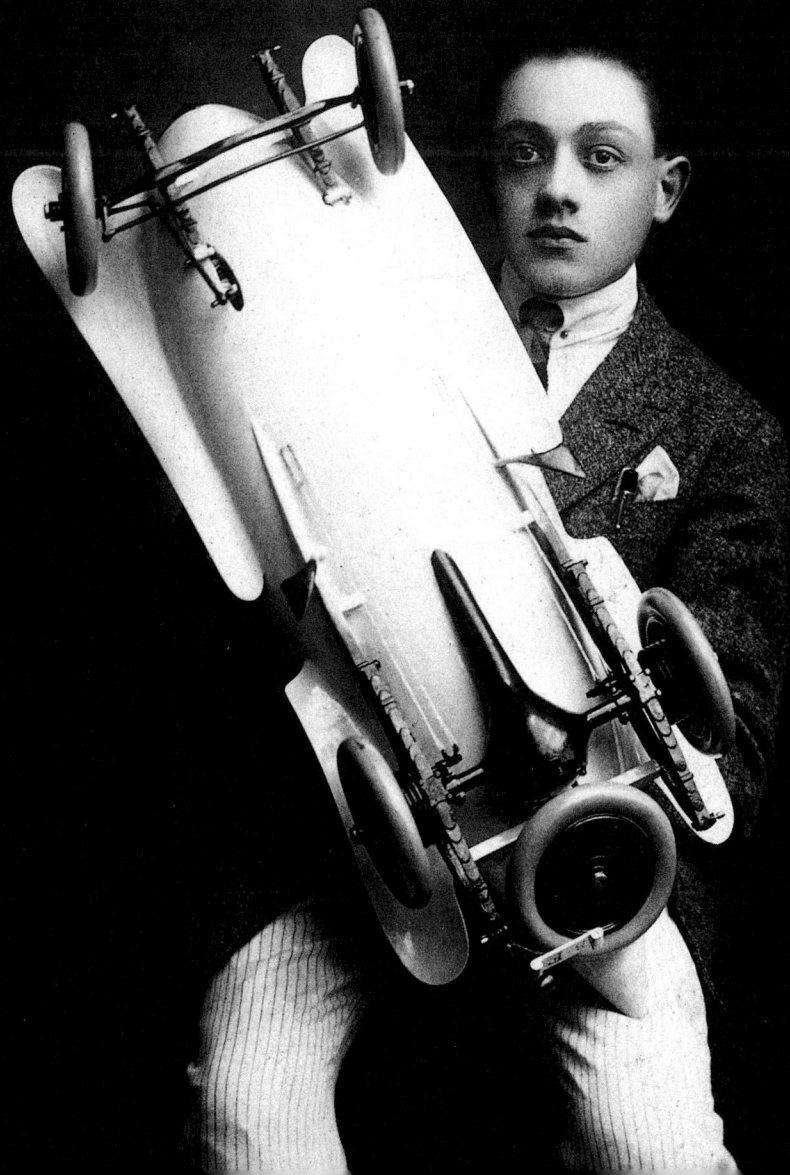

Flaminio Bertoni (1903–1964)

In 2001, the author of this essay edited a reprint of Alison Smithson's *AS in DS* through Lars Müller Publishers,[1] and in 2014 curated a small exhibition on the Citroën DS. The essay is part of an ongoing study with the working title "Fragments of the modern. Case studies, a design investigation."[2]

Form is the visible shape of content[3]

We align ourselves with this assertion by Ben Shahn when we analyze the characteristics of the shape of the form, which provoked such fascination in 1955. We are interested in the immediate effect and affect of the car's form, an example of the idea expressed by Frank Stella in his statements, *What you see is what you see* or *what can be seen there is there*.[4]

By referencing specific precepts developed by Gestalt psychology,[5] we will analyze the side, front and rear views of a blue Citroën ID (1959) and a red Citroën DS (1973). The cheaper ID model of 1959 is basically identical to the first edition of the DS from 1955; the DS model of 1973 is basically identical to that of 1967.

This study is concerned with the relationship of the parts to one another, the typical contour lines and, to the extent possible, the derivation of both from technical parameters. The analysis will not consider how the car is perceived in motion. There are many references to dynamic perception in advertisements.

1 – Alison Smithson, *AS in DS: An Eye on the Road,* reprint 2001, Christian Sumi (ed.), Lars Müller Publishers, Baden 2001.
2 – See bibliography in the appendix.
3 – In: Rudolf Arnheim, *Art and Visual Perception. A Psychology of the Creative Eye,* University of California Press, Berkeley 1954/1974, p. 96.
4 – Frank Stella, quoted in *Life* magazine vol. 64, no. 3, January 19, 1968, p. 49: "My painting is based on the fact that only what can be seen there *is* there. It is really an object. Any painting is an object."
5 – Gestalt psychology is concerned with the *immediate perception of form* in contrast to semiotics, which studies form "as a system of meanings." (Umberto Eco). Its central figure is Rudolf Arnheim and his book *Art and Visual Perceptions.* Rudolf Arnheim is the acolyte of Max Wertheimer, co-founder of Gestalt theory.

Flaminio Bertoni (1903–1964)

This way of seeing is based upon simple conceptual pairs: rising / falling, opening / closing, static / dynamic, front / back, above / below, expansion / contraction, flowing / disrupted, etc.[1]

The fundamentals of graphic analysis and the techniques of montage and collage are applied to a systematic series of photos by Michel Zumbrunn of both cars.[2] The analysis is further supported by technical drawings and sketches from earlier design iterations, advertising brochures, etc. Also considered, to the extent that they are relevant to the formal design, are the technical parameters set by the client, Pierre Boulanger, Citroën's director for many years, and his construction specialists.

The redesigned model 1967, first executed by Flaminio Bertoni in 1955, is a charismatic piece of design history, an impressive example of remodeling. It is also both an update to and radicalization of an existing form by means of minimal interventions (the front fenders are widened with "Cat's Eyes"). This fact begs the question of whether it is the same automobile or has become a different one. Roland Barthes dealt with this question. He points to the Argonauts who constantly reconfigured their space ship while in flight and, in true Structuralist fashion, he concludes that *they ended with an entirely new ship, without having to alter either its name or its form.*[3]

References to the "semantic content" of the form are found in the stylish advertising brochures produced by the most famous agencies and photographers of the time.

This study is not definitive. For that, it would require further studies in Bertoni and Citroën's archives. Instead, it has a fragmentary character. It attempts to push to their extreme a few arguments that address fundamental questions of form and its perception. The integrated series of images complement the analysis; at the same time, they are "autonomous" in and of themselves.

1 – See also Christian Sumi, "Forma e percezione: I televisori Doney, Algol e Black per Brionvega," in: *MZ Progetto integrato. Marco Zanuso design, tecnica e industria*, Alberto Bassi, Letizia Tedeschi (ed.), Archivo del Moderno (AdM), Mendrisio Academy Press / SilvanaEditorale, Milan 2013, pp. 54–65.
2 – The cars are now in the collections of the Verkehrshaus der Schweiz in Lucerne (blue ID 19) and the Museum of Modern Art in New York (red DS 23) as gifts of the author.

3 – *Roland Barthes by Roland Barthes*, Christian Sumi (ed.), University of California Press, Berkeley and Los Angeles, 1994, p. 46.

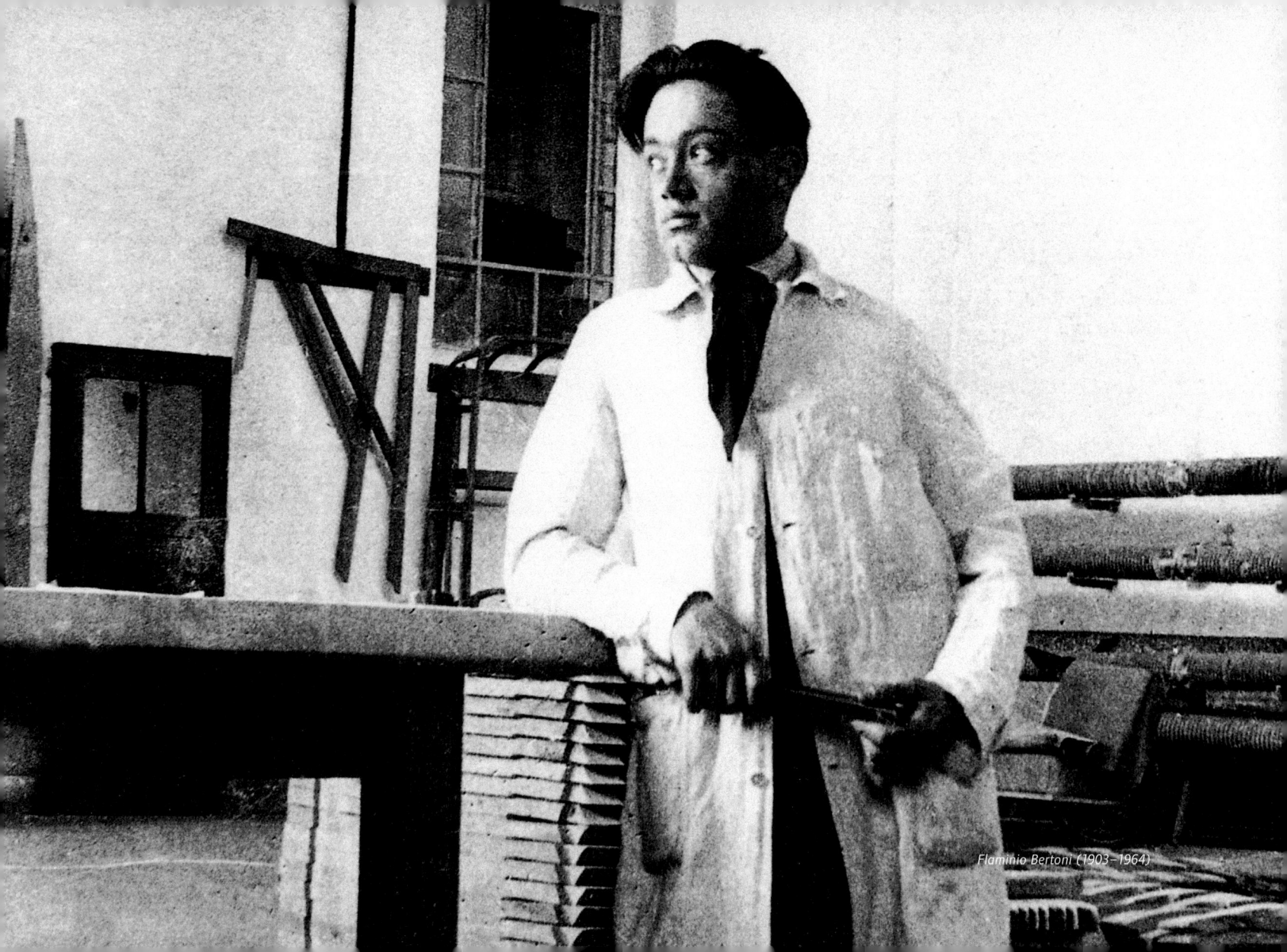

Flaminio Bertoni (1903–1964)

Citroën ID 19 and DS 23

Photos: Heinz Unger

Citroën ID 19 and DS 23 Photos: Heinz Unger

In different sequences, the photos depict the stringency and overwhelming beauty of the car designed by Flaminio Bertoni (1955), as well as the car's reconfiguration (1967). The DS is a design of the century within automotive design history. The photos illustrate in three dimensions, and in support of the analysis, how the contour lines of the side and front views were derived.

Blue: Citroën ID 19 (1959)
Red: Citroën DS 23 (1973)

Citroën ID 1959

Citroën ID 1959
Sketches[1] DS 1955

In 1938, only four years after the launch of the 1934 Traction Avant, studies begin for a new car. Sketches from Flaminio Bertoni's hand show side views of the Traction Avant with elegant, streamlined contours drawn on top of the car (a fluid transition from the radiator grille – front windshield – roof – rear), perhaps first attempts of the DS.

1 – Information can be found in, among other places:
– *DS toujours d'avant-garde / DS – Always Avant-Garde,* Shirine Guy, Antoine Demetz (ed.), Citroën communication, Gutenberg networks, France 2015
– Roger Brioult, *Citroën. L'histoire et les secrets de son bureau d'études,* Collection "Histoires d'Autos", n. 5, ediFree "La vie de l'auto," Fontainebleau 1987
The cheaper ID model of 1959 is basically identical to the first edition of the DS from 1955.

SOCIÉTÉ ANONYME
ANDRÉ CITRO
REPRODUCTION INTER

DÉSIGNATION SCHEMA DE L'ECHAPPEMENT

N° 309262

André Citroën had visited Henry Ford in America at the start of the 1930s and was thrilled by mass production and the production line (Fordism). The VGD Voiture de Grande Diffusion (or Mass Market Vehicle), if not the Traction Avant, is his answer to America's challenge.

One characteristic of further studies of the VGD after the war is the long, bow-shaped hood that lent the design its nickname "Hippopotame." The wheelbase between front and back wheels is long and the rear wheel is enclosed, as with the DS.

P 1134-53 .   Bertone

Later, the nose will be flatter and the fender ever more fully integrated into the overall form. The rear was elongated and descending, in accordance with streamline design; this resulted in too little headroom for the back seats. Only in 1955, the year in which the Citroën DS was introduced, did the design free itself from this limitation.

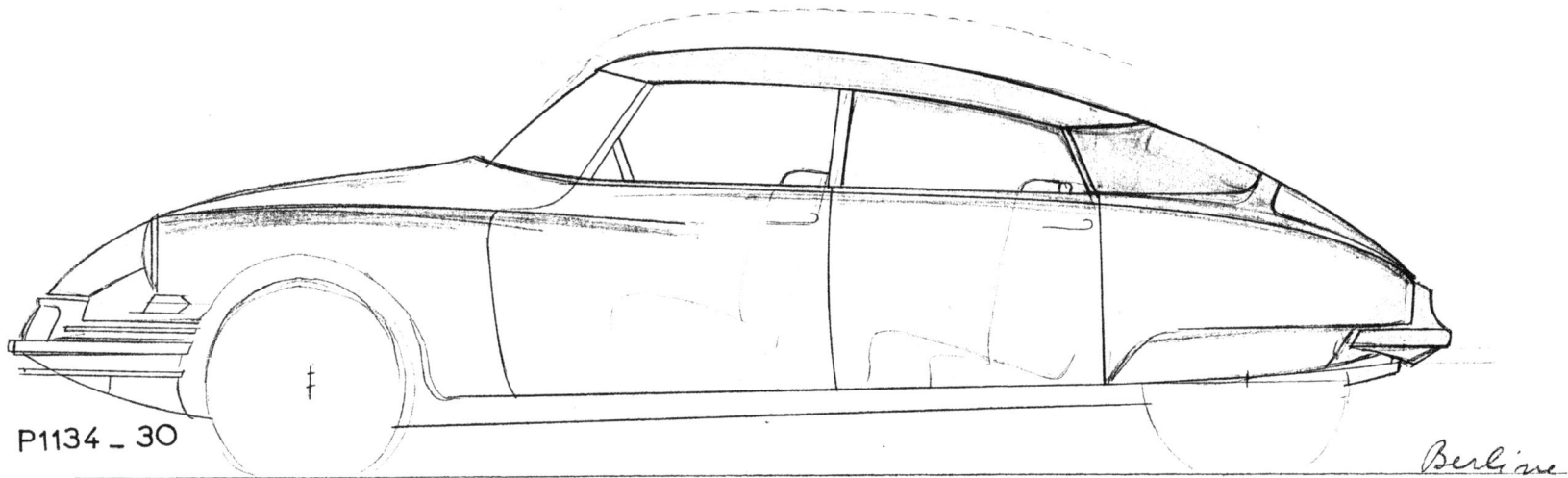

P1134 _ 30

Berline

P1231

The convertible assumed its final form a year before the sedan, as proven by a drawing from January 24, 1954.

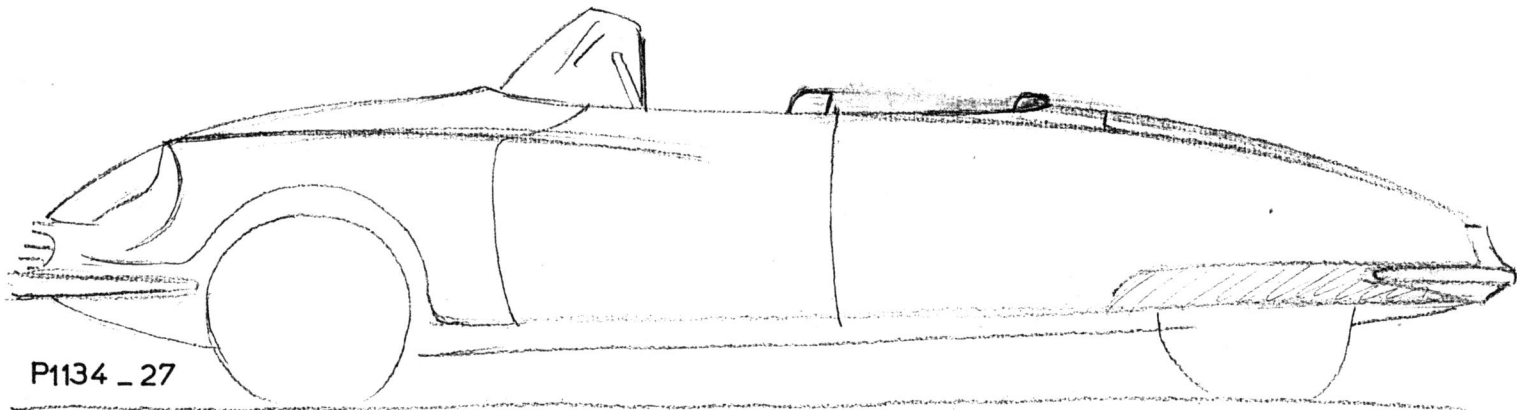

P1134 _ 27

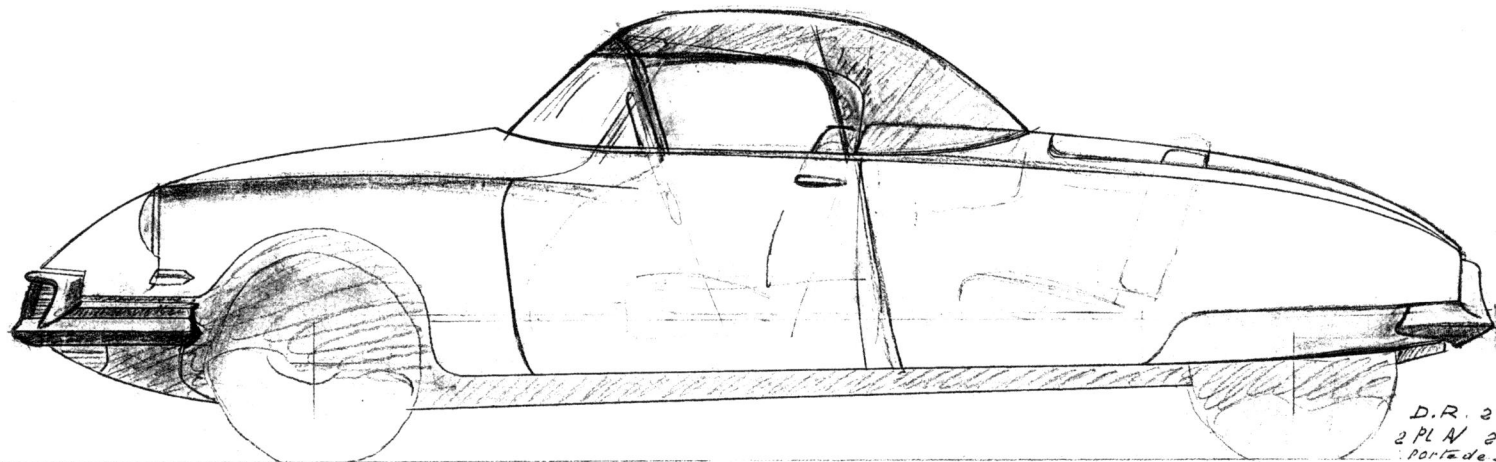

P1134 _16

D.R. 2 Porte,
2 PL AV 2 PL AR
Porte de serie
L = D serie

Citroën ID 1959
Analysis

Photos: Michel Zumbrunn

Side View
From "getting on" (a horse) to "getting in"

The side view is characterized by the increased wheelbase with the three columns A, B, C, as had already been the case with the Traction. This is particularly true of the partially enclosed rear wheel, which is pushed far towards the car's rear. This feature makes it easier to get into the rear seats and answers to Boulanger's requirements. The rear seats are positioned between the axels and not, as had been the case until then, above the rear axel. The seats "lay low," so that the passengers do not have to get "on," as if mounting a horse, but instead get "in." The disk brakes are positioned directly behind the motor ("tout dans le thorax") with a low center of gravity as André Lefèbvre, Citroën engineer, demanded.

C   B   A

3125 mm

"Pulling"
A horse does not push a wagon, it pulls it.

André Citroën on the front wheel drive of the Traction Avant 1934

The open, unenclosed front wheel and the semi-enclosed
rear wheel create a strong tension between "front and back"
(by contrast, see the open rear wheel of the DS Break) and
emphasize the direction of travel. The open front wheel drive,
in comparison to the enclosed rear wheel (for tire changes,
the entire rear fender had to be removed) recalls the side
view of a small airplane, with its two front wheels and small
rear wheel. The differentiation between open front wheel
and semi-enclosed rear conveys the impression that the car
is being pulled.

• Citroën DS Break 1970
• *See chapter: Advertising, pp. 186–187*

"Pushing"

The blinkers (also called the "trumpets of Jericho"), positioned high on the car's roof, are like boosters that "propel" the car (comparable to the Caravelle's rear-mounted engines, positioned on either side of the fuselage), in contrast to the open front wheels that pull it. The two opposing impressions create tension, an internal dynamic within the form.

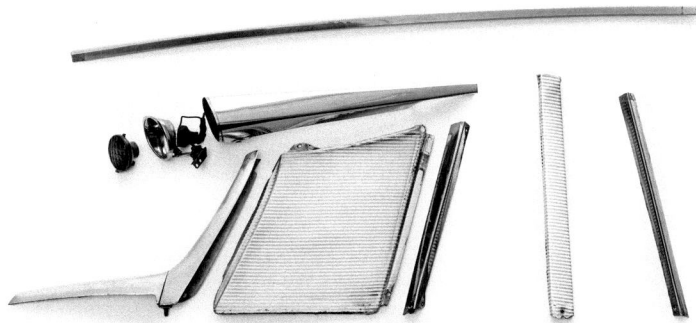

· Caravelle (1960)
· See chapter: Disassembly, pp. 100–101

"Stretching I"

The hood of the VGD, Citroën's DS predecessor, protrudes beyond the front wheel and lends the side view a "lateral elongation." Lefèbvre, the engineer, wanted the car to be heavier in front. For this reason, the motor[1] is positioned in front of the front wheel drive axel, in analogy to the 2CV. The symmetrically placed doors also correspond to early attempts at the 2CV. The form seems heavy-handed and bullish, which earned it the nickname "Hippopotame" (hippopotamus).

1 – See Presnell, op. cit., pp. 34/35, referring to the photo with the two convexities on each side of the two front cylinder heads, probably a flat box motor. Also see Fabien Sabatès and Didier Lainé, op. cit., Jacques Borgé Nicolas Viasnoff, op. cit.

· Citroën 2CV (1938) / Motor of a Citroën 2CV
· VGD after 1950: The new front with two convexities probably of the cylinder heads / Prototype flat box motor

P 113H-53

VGD (Voiture de grande diffusion)
As of 1939. Drawing from the period after 1945.

Purely on account of time constraints, this concept was abandoned for the DS, and the motor – a reworked version of the Traction Avant – was placed conventionally, behind the front wheel drive axel. The car's nose was flattened and the reserve tire placed in the "hollow" in front of the radiator, where the tire also served as a collision buffer. In combination with the long wheelbase and the low center of gravity, these features meant that the form was pulled and stretched in length.

Radiator grille with reserve tire

"Folding"

Folding the sheet metal over the chassis' longitudinal support line consolidated the rear fender, the doors and the front bumper, including its apron. In the case of the cabriolet model, a decorative molding underscores this effect (see page 135). The subtle bend beneath the windows in the car's upper section integrates the trunk and tail into the contour of the car's side view. The way in which the passenger cabin is enveloped "from rear to front" is even more evident in the convertible, with its flat trunk.

Citroën ID 19

Continuous contour lines

The transition between hood and fenders/doors renders a soft, rising contour that culminates in the A-column at the windshield, easily visible in the photo showing the disassembled sides. The front doors adjoin the fenders far at the car's front so that the traditional relationship to the A-column is "decoupled." The result is an overlap among the elements of doors, fenders, hood and bent windshield. The descending contour from front to rear end creates a counter movement, and the overlap of countervailing tendencies creates again an internal formal tension.

See chapter: Disassembly, pp. 98–99

"Ascending" "Descending"

Equally interesting is the genesis of the C-column at the rear window and the simultaneous integration of tail and roof in the form of the passenger cabin. An early sketch, in which the front already corresponds largely to its definitive form, depicts roof and trunk as connected to each other by a long, descending line. In a later sketch, the C-column is enlarged and shifted back; it "straightens itself up" and is "released" from the streamline form's dictates. C-column and roof now create their own figure, the bent rear window is "isolated" and in optical terms, the roof is pushed upwards as if it were a spoiler. In contrast to the "descending" silhouette shown in the first sketch, this solution is characterized by a counter movement, an "ascension" that is already apparent in an early wooden model from around 1950.[1]

1 – Dating according to *DS toujours d'avant-garde / DS – Always Avant-Garde,* op. cit.

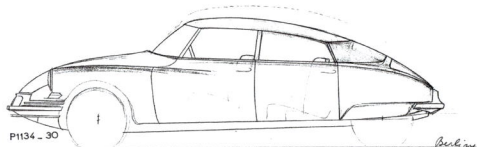

In the definitive version, the blinkers (in the case of the ID, these took the form of a red Bakelite funnel) are positioned between the C-column and roof. The C-column is now realized in a different material than the roof and asserts itself as an autonomous element between trunk and roof. The strongly expressive hinge on the roof, connected to the C-column, functions as an "optical" support for the part of the car body that sits on the trunk. The marked, stirrup-like brackets on the roof isolate the curved rear window, whose contour follows the descending streamline. Two new tendencies, two new figures overlap, one ascending and one descending. To some extent, the blinkers near the roof confuse the clear contours, especially in the case of the ID's Bakelite variant. Nonetheless, the blinker would become the car's emblem.

P1134 - 30

The overlapped pictograms summarize in abstract form the characteristics of the side view and its basic skeleton and contour lines. An illustrated example of a dynamic form abstracted to its basic structural skeleton is also found in Wassily Kandinsky's analysis of the dancer Palucca from 1926.[1] Concerning the theme of "dynamics" and its contour lines we should also mention *Danzatrice* (the dancer) by Flaminio Bertoni.

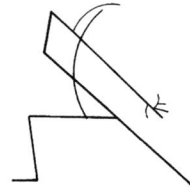

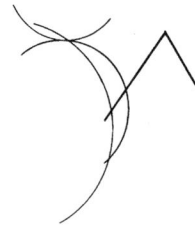

1 – Wassily Kandinsky, "Tanzkurven: Zu den Tänzen der Palucca," *Das Kunstblatt*, Potsdam, Vol. 10 (1926) no. 3, pp. 117–121. Thanks to Martin Steinman for this reference.

· Flaminio Bertoni, *Danzatrice* (1953)
· Wassily Kandinsky, "Tanzkurven: Zu den Tänzen der Palucca" (1926)

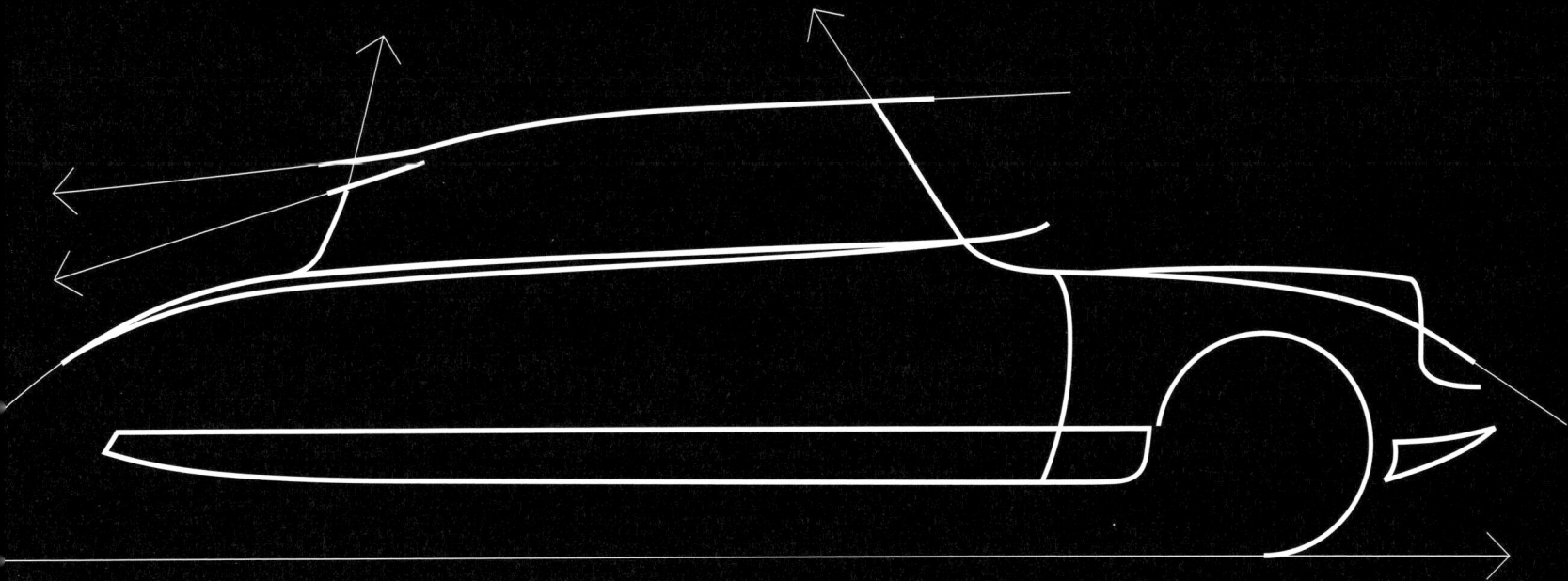

Front View
Beneath/above the bumper

The radiator's aeration vents are, perhaps in response to Lefèbvre's wish, located beneath the bumper at the motor's center, to the side of the engine-mounted disk brakes. The access point for the "hand crank" is also below the bumper. In this manner, the element that classically determines every car's appearance – its vertical grille – has disappeared from its front. It has been "banished" to the car's underside. The vents beneath the bumper recall the hungry mouth of a shark. Only the air vents for the passenger cabin are still above the bumper, or more precisely, are integrated into it. The car is no longer earthbound, but instead lifts off and seems to fly.

Aeration vents for the motor and the two disk brakes

Unmediated connections and integration

The hood is connected directly to the windshield across the car's entire width without an intermediary element. Curved windshield and sheet steel hood comprise a larger unit in geometric terms. In this way, two different materials are "transposed to a common form" without mediating elements.

From "pushing" to "connecting"

The lower front apron with its air vents is the same color as the hood and thus becomes part of the car's body. The bumper is extended to the cut-out around the front wheel and conjoins "above" and "below." In other words, the bumper no longer "bumps" but instead conjoins, as if it were a seam. The circular bumper lends enormous tension to the radiator face, and works as a conjoining element between front apron and hood.

Rear View
"Tapering" from front to back

The car's front track is broader than its rear, perhaps again because of engineering requirements related to vehicle path. In plan, this defines the typical drop form, the streamline form. As a termination to the form, the car's rear, box-like bracket repeats "at small scale" the squared form of the trunk, clearly seen in the convertible's elongated tail.

Citroën DS convertible

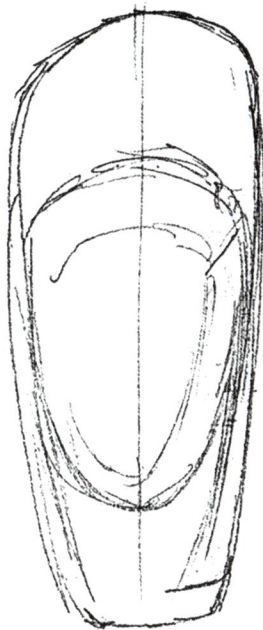
Sketch by Flaminio Bertoni (1954)

Front View / Rear View
"Framing"

The dimension of the bracket around taillights and license plate corresponds to the dimension of the front bracket around the air vents at the radiator. In the ID, the brackets that are part of the front and rear bumpers are more or less the same size, assembled from different pieces.
The comparison refers to the dimension of the bumper and not to the overall height of the car.

"Opening" "closing," line and joint

The lines drawn by the joints between fender and radiator cover differ from the lines drawn between the trunk and rear fender. The front line widens, "opening" a dynamic whereas the rear line narrows, "closing" the dynamic. The striking headlights, pushed towards the car's exterior edges, contrast to the small cat eyes set closer together near the trunk. The headlights augment the formal dynamic in the direction of travel. This effect is increased by the omission of the apron at the car's tail relative to its front.

Citroën DS 1973

Citroën DS 1973
Sketches[1] DS 1967

In 1962, the reconfiguration of the DS's front is a topic in the Citroën design studio, in particular the integration of high beams and low beams into a single element. By way of a predecessor to this idea, it is important to mention Bertoni's drawings from 1954. The headlights have already been completely integrated into the fender, in other words, into the car body.

1 – Information can be found in, among other places:
– *DS toujours d'avant-garde / DS – Always Avant-Garde*, Shirine Guy, Antoine Demetz (ed.), Citroën communication, Gutenberg networks, France 2015
– Roger Brioult, *Citroën. L'histoire et les secrets de son bureau d'études*, Collection "Histoires d'Autos", n. 5, ediFree "La vie de l'auto," Fontainebleau 1987
The DS model of 1973 is basically identical to that of 1967.

P.1204_3   ℬ. 2-12.54

P.1204_10   ℬ. 2-12.54

The following designers participated in the revised design[1]: Robert Opron, who succeeded Flaminio Bertoni after his death in 1964; Henri Dargent, Bertoni's assistant; Michel Harmand; Jean Giret; Regis Gromik; and Jacques Charreton.

Henri Dargent's designs from April 1962 show the first revision of the nose. High and low beams are already unified behind a glass lens. Both fenders are slightly widened, as in the definitive version.

1 – Information can be found in: *DS toujours d'avant-garde,* op. cit.

Henri Dargent (1962)

Robert Opron's designs (drawings from November 12, 1962) intend a striking change to the entire car body. The hard lines and the rear form point towards the later CX and SM, both by Opron.

Robert Opron (1962)

Opron's design (from January 1964) is now more reserved: the headlights are placed within the car body as an elongated rectangle.

P. 4302.4

Robert Opron (1964)

In the designs by Michel Harmand from July 1964, the high and low beams are arranged vertically.

Michel Harmand (1964)

452 D

P5013

Another study by Michel Harmand from 1966 shows the horizontal arrangement of three lights (as was done later in the SM) with significantly wider fenders.

The side view from 1967, again by Harmand, corresponds in most aspects with the final version of the DS.

P5484-?

Michel Harmand (1966)

P_5719_1

Michel Harmand (1967)

Citroën DS 1973
Analysis
Photos: Michel Zumbrunn

Side View
"Stretching II"

Compared to the older version, the new headlights (high and low beams consolidated behind a single lens) extend the fender to the height of the hood's front edge, or the height of the bumper (see disassembly photo). The two fenders end at the same height as the hood and meld to a single form. By virtue of the angle at which the fender's front edge is cut, the headlights are also visible from the side, lending the form additional dynamism. In other words, the new headlights influence not only the car's front view but also its side view. In the process, they become part of the overall form.

See chapter: Disassembly, pp. 98–99

Rubber extrusions placed beneath the bumper lend the car even more momentum. Together with the front apron, they make the impression of a protruding lower jaw. The flatter, compressed silhouette of the older model's fender interrupts the car's contour and is, by comparison, autonomous in relation to the hood.

Front View
"Integrating" "dynamizing"

The headlights disappear behind a kind of tiny windshield. The glazing is flush, integrated into the car's body and becomes part of its encompassing surface. The front fender is slightly broader, its gentle upward bend abandoned in this model, and the hood narrows towards the front. In comparison to the older model, the headlights are lower, giving the front a more "compressed" appearance.

The height of the vent integrated into the bumper is reduced even more, until it is little more than a slit. In the process, the car's hood is elongated. The tapering and elongation of the radiator cover makes the front even more dynamic. At this point, license plates and rubber extrusions are also located beneath the bumper.

See chapter: Disassembly, pp. 96–97

"Cat's Eyes"

The new headlights set themselves apart from the classical connotation of round car headlights and become movable cat's eyes. The round form of classical headlights is replaced by an elongated, conical form and in turn determines the form of the fender – or perhaps it is the other way around. In contrast to round, stable, peripherally-located headlights, the new model's headlights are orientated to the center. The blinker's form becomes more striking and more pronounced, in contrast to the new convex door handles.

In the SM, the "radiator motif" and its bracket are integrated into a "light muzzle" with the licence plate. A similar solution can be found in the Lamborghini Marzal (1967).

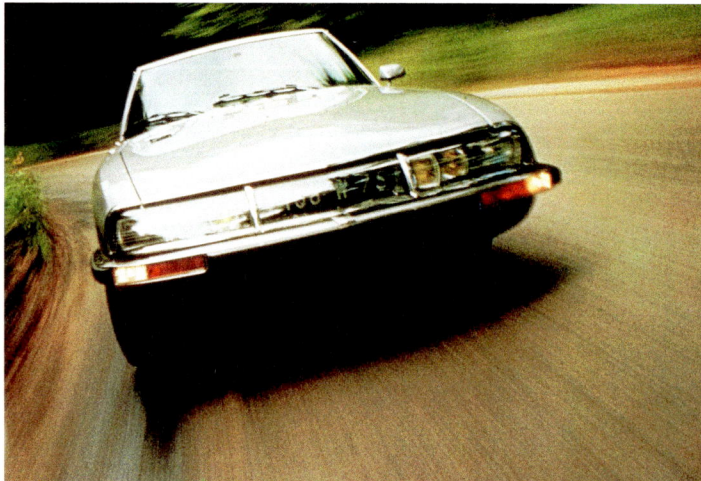

Citroën SM (1970)

How the swivelling lamps work

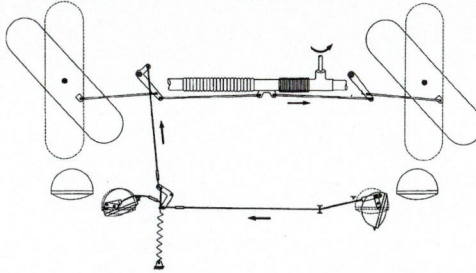

There is nothing complex about how the directional lamps work: it is just a system of spring-loaded rods and levers. Note how on full lock the inner lamp pivots more acutely, directing maximum illumination at the inside of the bend.

The larger outboard lamps do not turn, although they retain as part of the package the self-levelling mechanism introduced in 1965 on the DS21. It is only the smaller inboard lamps that pivot; these are long-distance driving lights with iodine bulbs and operate only on full beam.

The linkage to the steering rack is by a system of spring-loaded rods and levers and the axis of rotation is slightly off the vertical to compensate for body lean when cornering.

The clever thinking behind the lamps is not just that they turn as the steering wheel is turned, but that as a consequence of this, they turn a fraction before the vehicle itself begins to change course. This is because in cornering there is always a slight delay before the turning of the wheel is translated into a turning movement at the tyre contact patch. As a result, the pivoting lamps light the corner a little before the car starts to turn.

Typical of Citroën ingenuity is that the linkage ensures that the inner lamp always pivots more acutely than the outer lamp, with the difference becoming more exaggerated as the lock is increased: on full lock the inner lamp turns through 80 degrees. This means that maximum illumination is directed at the inside of the bend.

Similar swivelling and self-levelling systems were later used on the SM, but did not find their way into the CX.

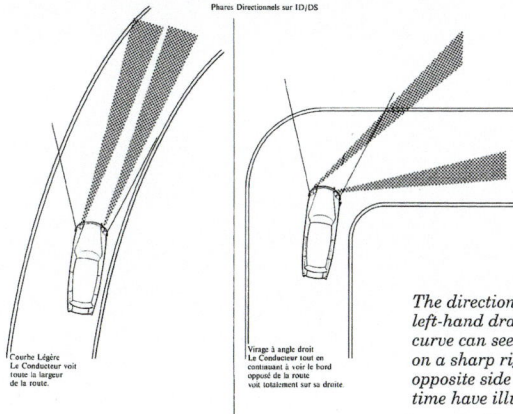

Phares Directionnels sur ID/DS

Courbe Légère
Le Conducteur voit
route la largeur
de la route.

Virage à angle droit
Le Conducteur tout en
continuant à voir le bord
opposé de la route
voit totalement sur sa droite.

The directional lamps explained. In the left-hand drawing the driver taking a gentle curve can see the entire width of the road; on a sharp right-hander he can see both the opposite side of the road and at the same time have illumination around the bend.

· *Extract from book: Jon Presnell, "Citroën DS," p. 87*
· *Citroën DS (1967), Advertising*

Rear View

In opposition to the front view, there are practically no changes between the 1959 model and the 1973 model.

Front View / Rear View

In opposition to the 1959 model, the bumper along the
radiator's front is, unlike the rear bumper, compressed,
creating a "front/back" dynamic.
The comparison refers to the dimension of the bumper
and not to the overall height of the car.

Disassembly Citroën DS 21

Photos: Heinz Unger

Disassembly Citroën DS 21[1] Photos: Heinz Unger

An advertisement from 1959 shows a family in front of a disassembled DS and its empty aluminum skeleton. This was the inspiration to deconstruct a Citroën DS 21 into its components and to photograph the process systematically. The chronology of disassembly brings to the fore the car's striking overall lines, while the single tableaus show the body's individual determinative components.

1 – Executed by students of the AAM Mendrisio at Garage Haefliger und Kunz AG: Simone Biaggi, Ladina Danuser, Edoardo Ferrari, Marco Magnani, Alina Maksutova, Martino Pedroli, Roberta Poretti, Luana Rossi, Mihran Rovelli, Sofiya Sayfullina, Artem Spiridonov. Assistant: Gianluca Gelmini.

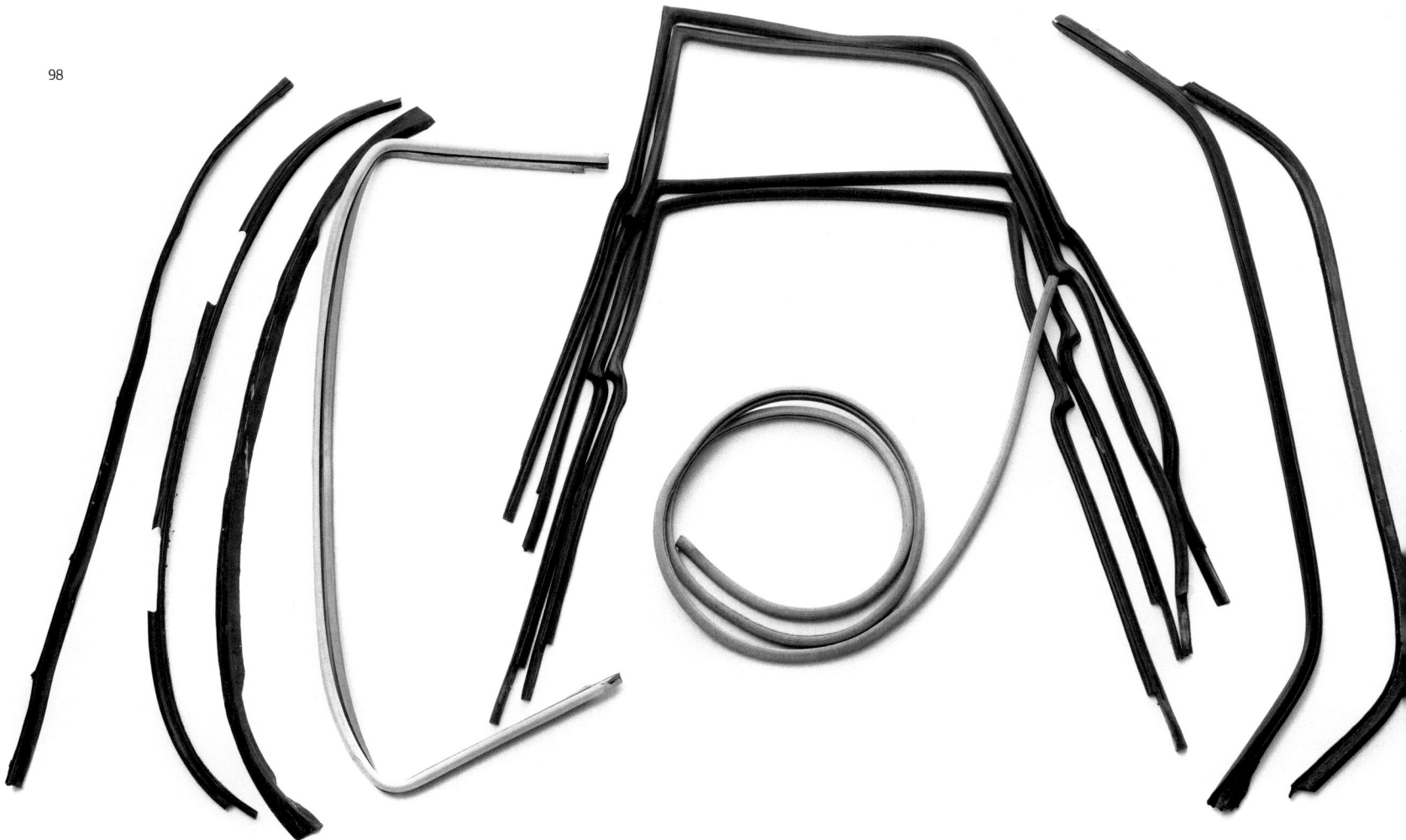

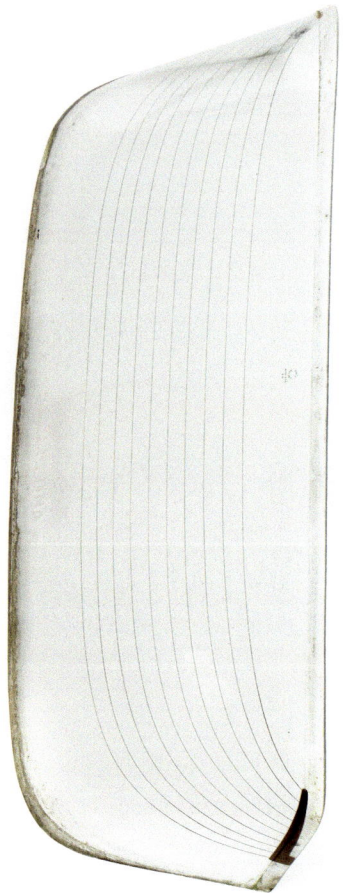

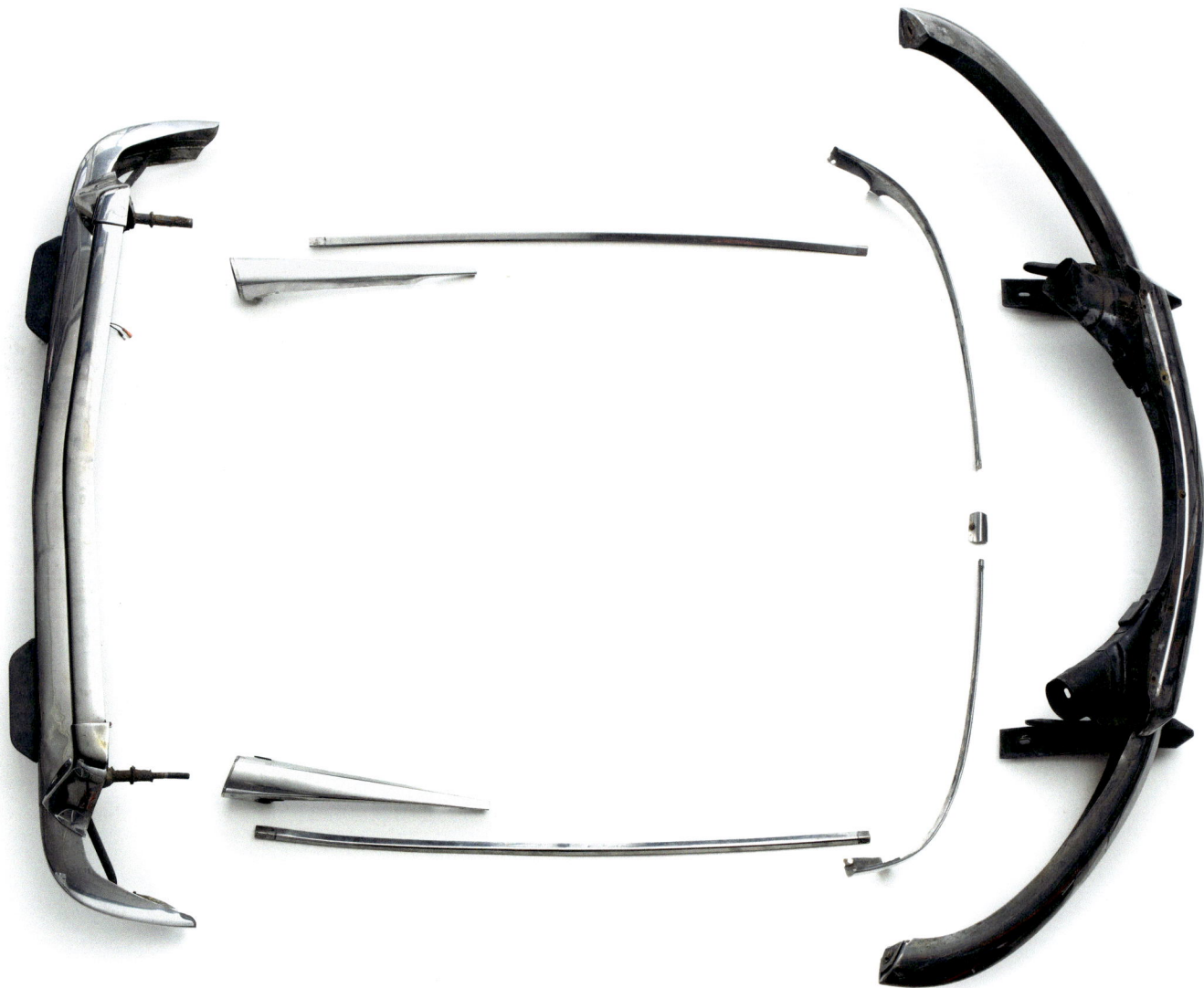

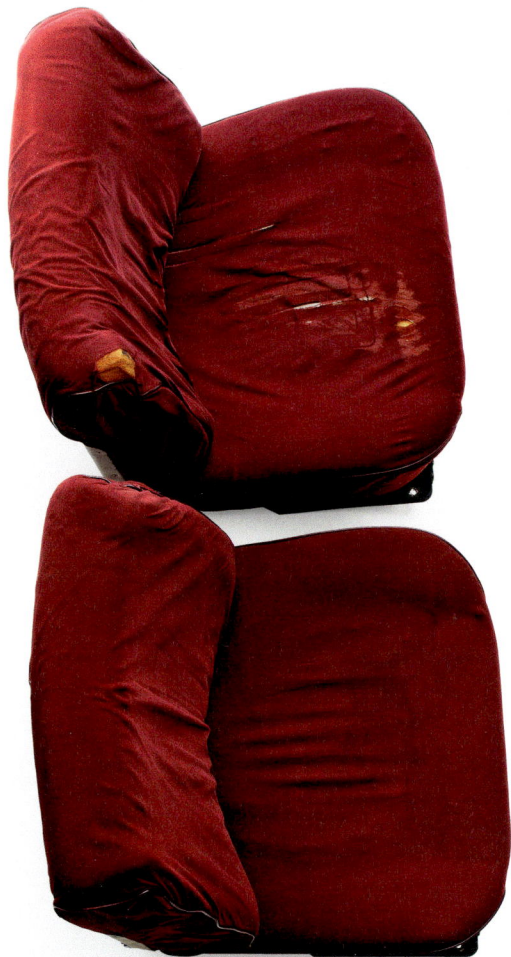

Synthesis

Synthesis

Rudolf Arnheim / Roland Barthes / Max Bill / Flaminio Bertoni / Otl Aicher / Gio Ponti

Raymond Loewy

For a moment, it makes sense to look back to a car that was produced two years earlier and was equally decisive for its style, the Studebaker Commander Starliner (1953) by Raymond Loewy. Discussing the car's genesis against the background of Loewy's design, the automobile journalist Robert Cumberford wrote, *The styling consists of the 1953 Studebaker sedan's lines superimposed on the architecture of the 1934 Citroën.*[1]

1 – Robert Cumberford, *Auto Legends*, op. cit., p. 179.

· *Flaminio Bertoni: Traction Avant (1934)*
· *Raymond Loewy: Studebaker Commander Starliner (1953)*

Flaminio Bertoni: Citroën ID 19 (1959)

The sculptural quality of the car nose in Loewy's studies for the new Studebaker and those by Bertoni for the DS have astonishing similarities.[1] Both cars repress the radiator face and by doing so, "destroy" the classical radiator motif.
The round "rocket booster" motif in Loewy's car refers to the Studebaker Commander Land Cruiser (1951); it is entirely eliminated two years later. The degree to which Bertoni was aware of Loewy's work is unknown.

1 – These similarities extend to the side view and also include the striking C-column. Quite dissimilar, however, are the Studebaker's doors, which are cut along straight lines, whereas the DS doors are overlaid. In addition, in the case of the DS, the typical connective element between hood and windshield is omitted. The direct connection between operable hood and curved windshield is extremely difficult to achieve in technical terms.

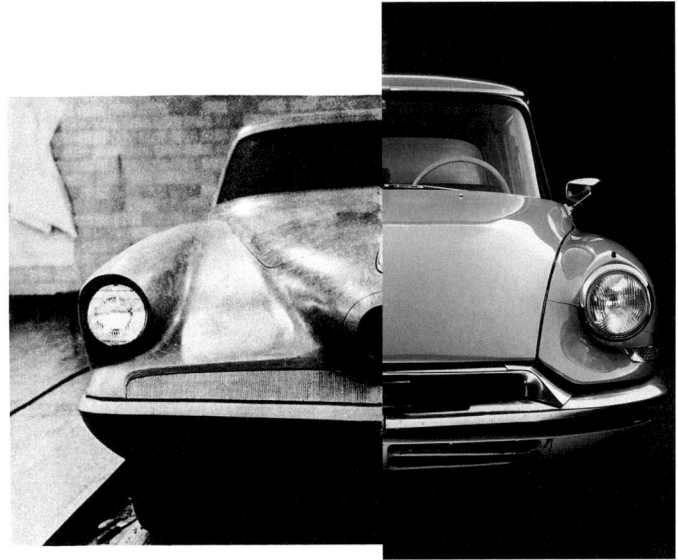

The radiator grille later returns in the Studebaker President Sky Hawk (1956), perhaps a concession to popular taste: a car without a radiator grille is not a "proper" car.[1]

Loewy's absolute innovation in the Studebaker Avanti (1962) was the use of fiberglass in the car body. The fin-shaped fenders have come to resemble thin guard rails, the headlights have been "banished" from the fender and are instead in the radiator housing. This car would be determinative for the style of the 1960s.

1 – The radiator, in a certain sense the car's breath, is today still a characteristic icon of various car manufacturers, such as, for example, the horseshoe-shaped radiator of the Bugatti (the manufacturer's founder was a horse lover); the filleted equilateral triangle on the Alpha Romeo; or the bipartite radiator, split in the middle, on the BMW.

· *Commander (1951)*
· *President Sky Hawk (1956)*

Studebaker Avanti (1962)

Raymond Loewy: Analytical drawings

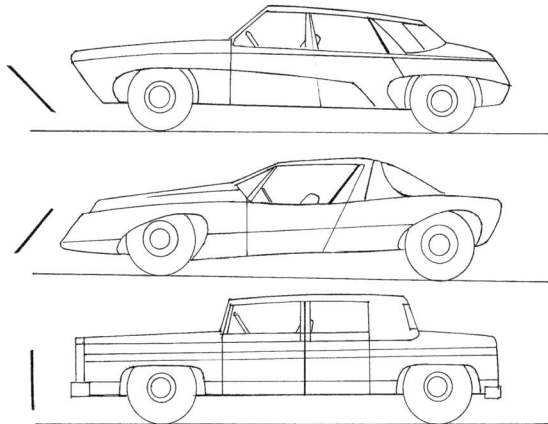

1. Homogenization of the Form
The Whole is Greater than the Sum of its Parts[1]

Dynamism

As early as the Traction Avant, Bertoni had transformed the appearance of the car body from the typical boxy form made by combining parts. By slanting the radiator housing and windows, he was able to overcome the stasis of the "box" and make the form dynamic, "silkier." The front fender, the underside of the low-hanging body and the rear fender form a coherent, "flowing" line. The body is now more than the sum of its parts, a tendency that will be further developed in the Citroën DS.

Raymond Loewy[2] and Otl Aicher[3] would later comment, by means of genealogies that they constructed, upon the dynamism that the form gained by angling the car's front (ascending) and rear (descending).

1 – Rudolf Arnheim, op. cit., p. 78.
2 – Raymond Loewy, op. cit., p. 197.
3 – Otl Aicher, op. cit., p. 73.

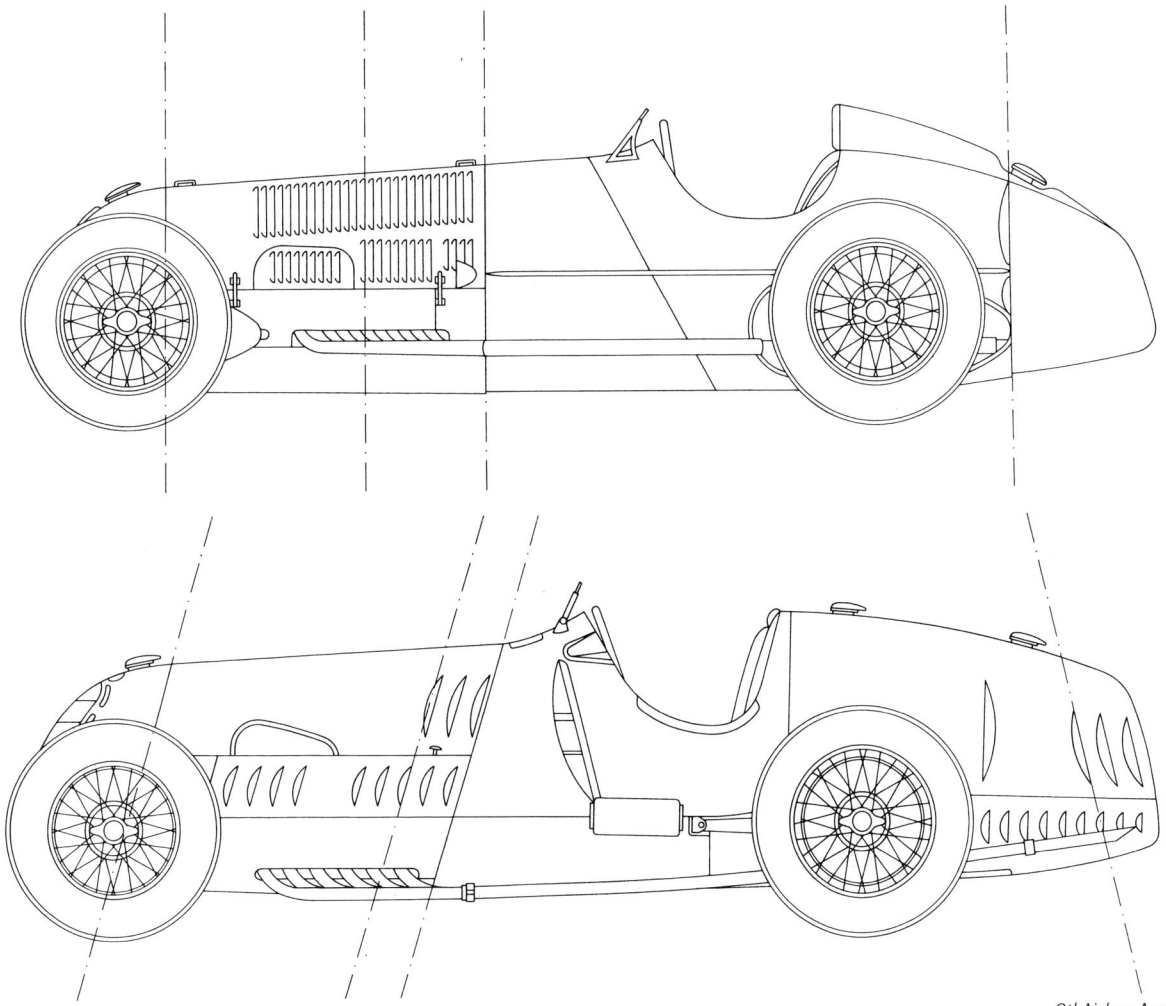

Otl Aicher: Analytical drawings

120

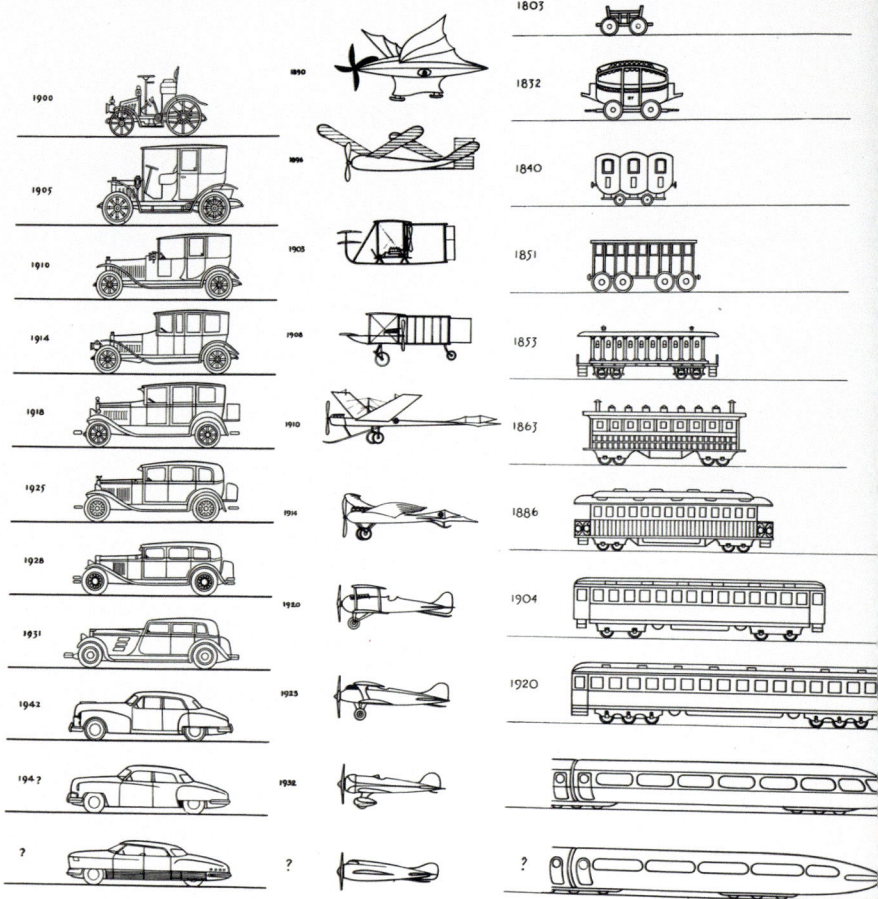

EVOLUTION CHART OF DESIGN — 1930

Extract from:
Raymond Loewy, "Industrial Design," pp. 74–76

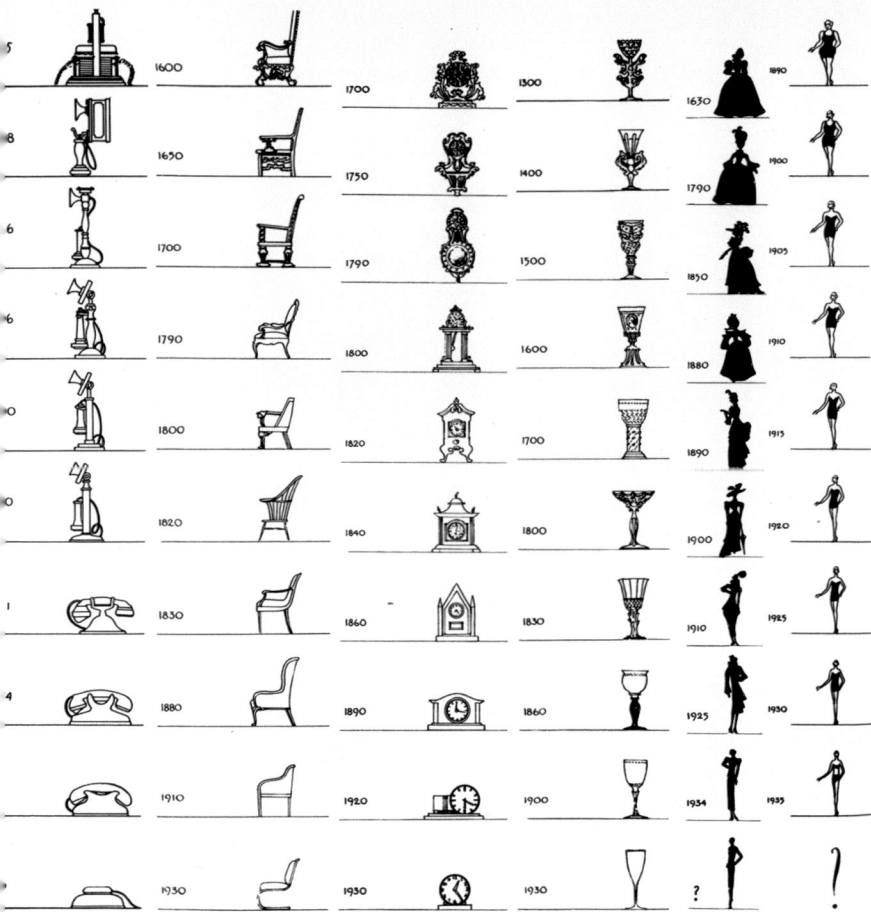

5	1600	1700	1300	1630	1890	
8	1650	1750	1400	1790	1900	
6	1700	1790	1500	1830	1905	
6	1790	1800	1600	1880	1910	
0	1800	1820	1700	1890	1913	
0	1820	1840	1800	1900	1920	
1	1830	1860	1830	1910	1925	
4	1880	1890	1860	1925	1930	
	1910	1920	1900	1934	1935	
	1930	1930	1930	?		!

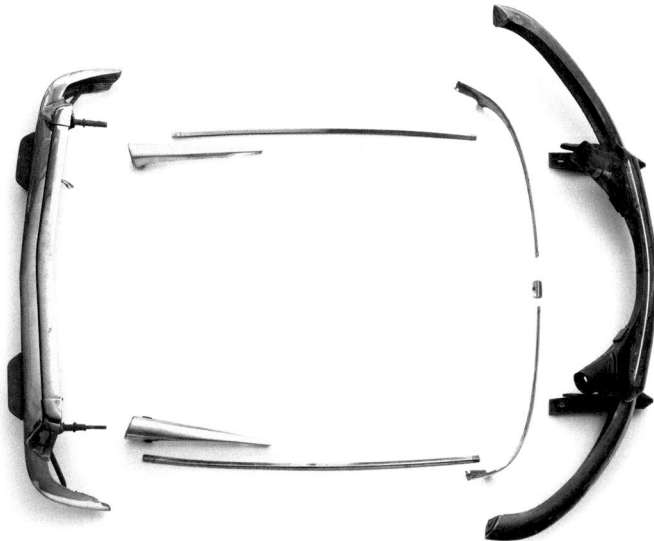

Unmediated joinery
A Whole Maintains Itself [1]

The body parts for the radiator housing and trunk align directly with the curved front and rear windshields of the DS. Rainwater is drained using small, u-shaped channels within the body. The seams create a system, a figure of continuous lines. These lines never meet at a single point but are shifted against one another. By avoiding intersections, the design allows the components to interlock.

At the same time, when viewed from above, the contour lines are "concentric," like a Russian Matryoshka doll: roof outline, upper and lower outlines of the car windows, the overall contour. In other words, the distribution of conjoined body parts is cohesive, as is the figure that results from the continuous seams, and its inscription within the larger form as a method of division – all of this results in the form's consolidation.

1 – Arnheim, op. cit., p. 67.

See chapter: Disassembly, pp. 104–105

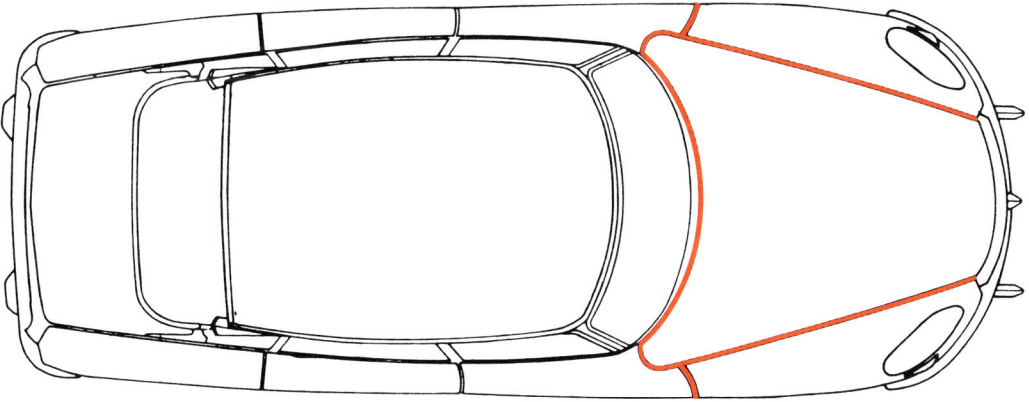

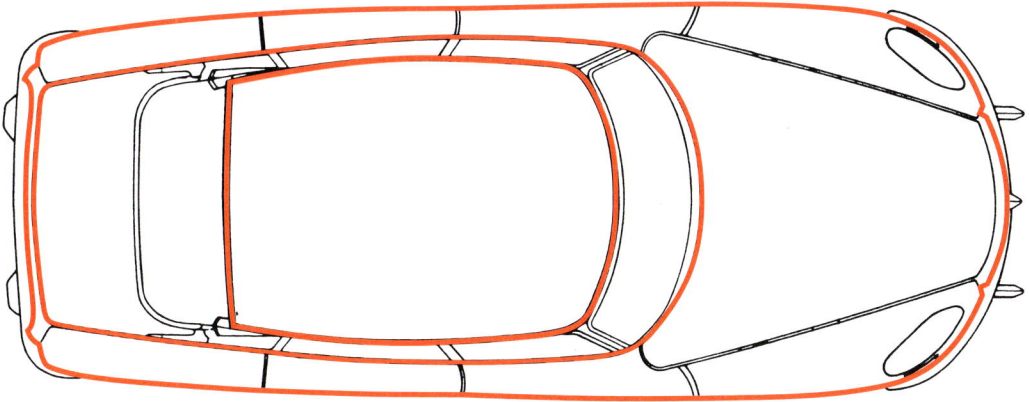

Lines of forces

The contour line of the body opens / pulls the car forward, the contour line of the cockpit closes / pulls the car backward.

The (almost circular) round bumper pushes itself forward like a plow, like the prow of a ship cutting the water.
It acts like a taut cable, juxtaposed to the car's forward motion while at the same time "illustrating" it. This is quite different from Loewy's Studebaker Champion Starliner, with its straight, slightly filleted bumper.

· *Flaminio Bertoni: Citroën ID (1959)*
· *Raymond Loewy: Studebaker Commander Starliner (1953)*

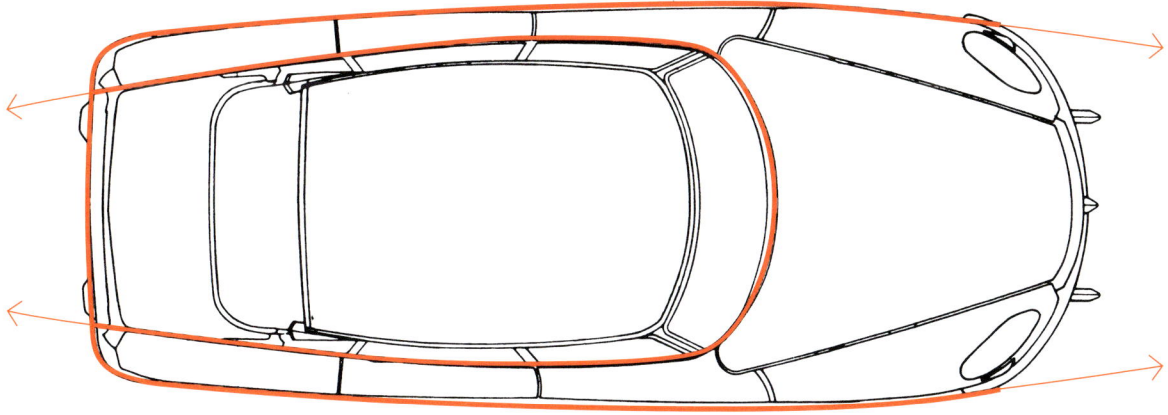

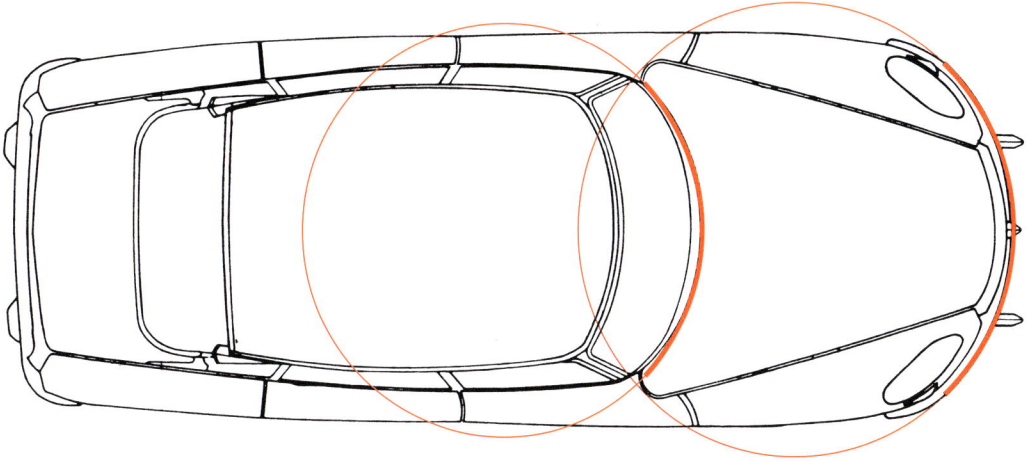

Assembling

The basis and the precondition for unmediated joinery among the body parts are the compression strips made from rubber or newly-introduced neoprene (connections, joinery of glass to metal or plastic components, moving parts such as doors). Of this phenomenon Roland Barthes wrote, *There is in the DS the beginnings of a new phenomenology of assembling, as if one progressed from a world where elements are welded to a world where they are juxtaposed and held together by sole virtue of their wondrous shape...*[1]

1 – Barthes, op. cit.

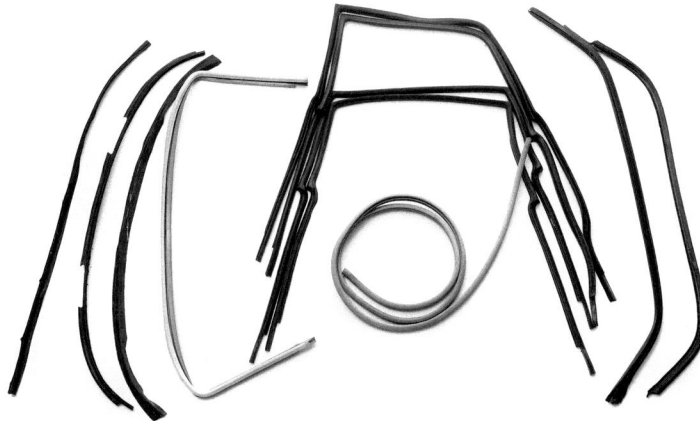

Welding see chapter: Disassembly, pp. 98–99

Model Citroën DS with closed fender

"Contour Commun"

Like the wine bottles, carafes and different glasses in the Purist images by Amédée Ozenfant[1], individual body parts and elements are "wed" (mariage d'objets): the hood, fender and doors of the DS form a dense composition. The seam in the form of a rubber membrane or little water channels becomes a shared contour or seam, in turn analogous to the shared contour lines among different objects in a Purist image (contour commun). The auto body parts are integrated into a compact form and "grouped" with one another.

1 – See Christian Sumi, *Immeuble Clarté Genf*, op. cit., p. 104.

· *Amédée Ozenfant: Puristic drawing "Contour commun" (1968)*
· *Le Corbusier: Illustration "Mariage d'objets" (1968)*

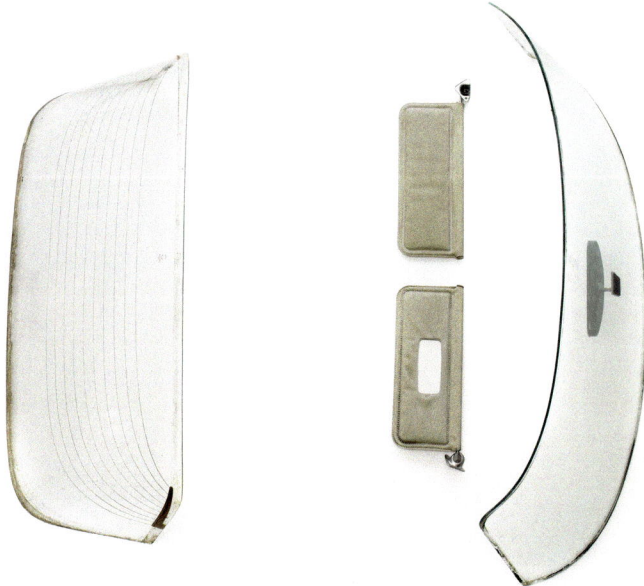

Increasing formal simplification
Reveling and sharpening[1]

With the reworking of the car's body, the form was reduced to overarching lines and its tendency toward "Prägnanz" was reinforced. This homogenization of form worked in favor of the pontoon carrosserie.

Glass and sheet steel form a unit, a whole. The curvature of the body parts' contour lines follows the large curved windows. The DS's glass surfaces comprise a total of 2.25 m² with a 360-degree panoramic view effectively staged in advertisements. Roland Barthes wrote, *The DS is obviously the exaltation of glass and the pressed metal is only a support for it. Here the glass surfaces are not windows, openings pierced in a dark shell, they are vast planes of air and space.*[2]

1 – Arnheim, op. cit., p. 66.
2 – Barthes, op. cit.

See chapter: Disassembly, pp. 102–103

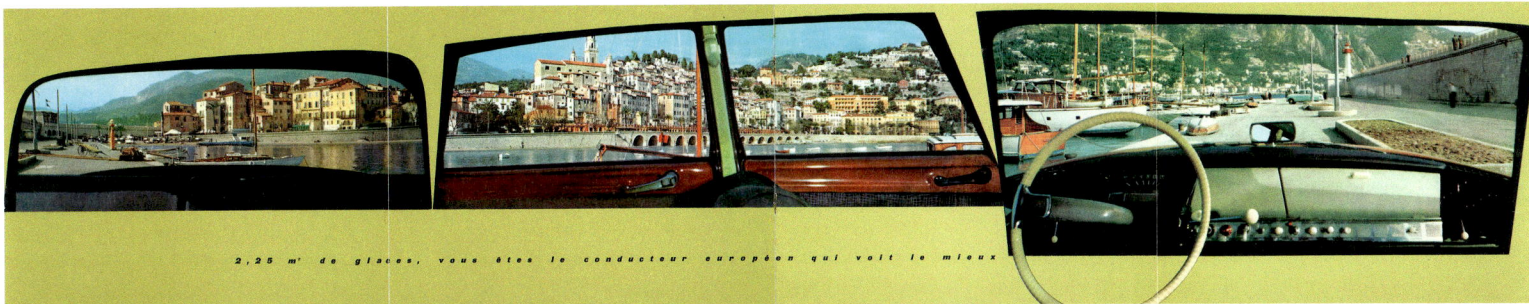

2,25 m² de glaces, vous êtes le conducteur européen qui voit le mieux

• Panorama view: Citroën DS/ID
• Panorama view: Raymond Loewy, Studebaker Champion Regal De Luxe (1950)

Break

The DS's step-like C-column seems to contradict the tendency described above towards the homogenization of form and with it, streamlining. However, an element that "ascends," "breaks away," fragments, frees itself from the overall form, creates additional tension. This is especially true by comparison with the reclining, often unexciting tail of classic streamlined cars (Jaray, Tatra, DKW), but also, by comparison, with Bertoni's early sketches for the DSs. Therefore the car's side view becomes more concise. Interestingly enough, this was a requirement of the manufacturer, who wanted more headroom.

In the case of the Ami 6 (1961) the expressively "inset" rear windshield and C-column mark the definitive end of the classic streamline thematic (see also later the convex rear windshield of the CX by Opron), and an entirely new design direction is set. The hood appears like a textile blanket laid on top of the motor, and the blanket's rolled edge becomes a curving radiator silhouette.

Mockup Ami 6 (1961)

Paul Jaray's own car (1933), streamlines drawn on it for demonstration purposes

Body and compartment (cockpit)

As to the relationship between roof and the body, the convertible designs are particularly instructive. Beginning at the front bumper, the elongated line runs along the top edge of the doors. The folding roof "glides" along this line. In other words, the body is an independent form and the cockpit an element placed atop the rest. According to the dated drawings, the car's convertible version is earlier, developed before the sedan (1954), which explains the difficult struggle to find a back-front tail form for the sedan.

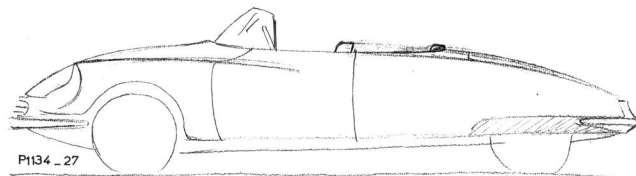

P1134_27

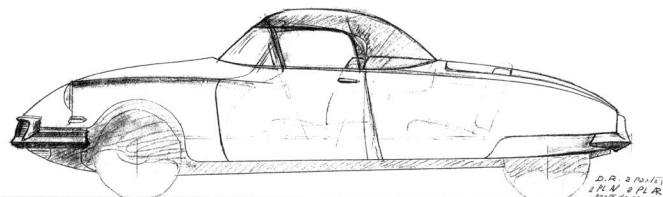

P1134_16

Citroën DS 21 convertible (1966)

Isomorphism[1]

When visual form and perception are in agreement – form as the structure of visually oriented external forces (visual form) and internal forces (perception) – Gestalt psychologists will speak in terms of isomorphism. The precondition is that the form is strong and impressive enough to remain "suspended" in memory, a characteristic that accrues to the DS, as we have tried to demonstrate.

Roland Barthes writes, *The DS is first and foremost a new Nautilus, one of those objects from another universe.*[2] The DS's form does not reference "driving," it references "floating," a fact that in turn supports its many potential readings.

1 – Ibid, p. 63.
2 – Barthes, op. cit.

See chapter: Advertising, pp. 172–173
Robert Dumoulin (1959): Flying car with family

Nous avons réuni à votre intention quelques
extraits d'articles de presse consacrés à la
présence de la DS 19 à la TRIENNALE DE
MILAN. En guise de préface à cette gerbe
d'opinions, voici un court passage d'une
allocution prononcée par le célèbre styliste
italien GIO PONTI lors de sa réception à
PARIS, en janvier 1957, par l'Institut d'Esthé-
tique Industrielle :

"... Je veux dire l'admiration que nous

autres, italiens, nous éprouvons pour la

dernière-née des voitures françaises, la

DS 19. Cette voiture a le courage d'être

une machine sincère. Elle ne cherche pas,

comme les productions de l'École amé-

ricaine, à séduire l'acheteur, par des

barbouillages multicolores terribles, des

chromes abondants, un effort pour tout

masquer. L'École européenne, elle, écoute

la technique ".

Robert Dumoulin: Advertising with quote from Gio Ponti (1958)
The left headlight is not visible and therefore the flying car is more dynamic.

· *Traction Avant (1934)*
· *2CV (1949)*
· *DS (1955)*
· *Ami 6 (1961)*

Hybrids

The 1967 adjustments to the car's front recharged the "visual form and its perception." The attempts to combine Bertoni's different models, to "hybridize," for example, by placing the radiator housing of the Traction Avant onto the 2CV, or that of the DS onto the Ami 6, remain nothing more than curiosities. The "failure" of these attempts – although the hybrid DS / Ami 6 with its combined headlights has a light and cheeky quality – points to the cars' individual characters, to the singularity of Bertoni's four designs. They are four fundamentally different conceptions with different qualities. Unlike the designs of, perhaps, Pininfarina, which are characterized by a certain continuity of themes and lines, each Bertoni design begins anew. In this sense, his work is marked more by "breaks" than by "continuities."

· Traction Avant / Citroën 2CV
· Citroën DS / Ami 6

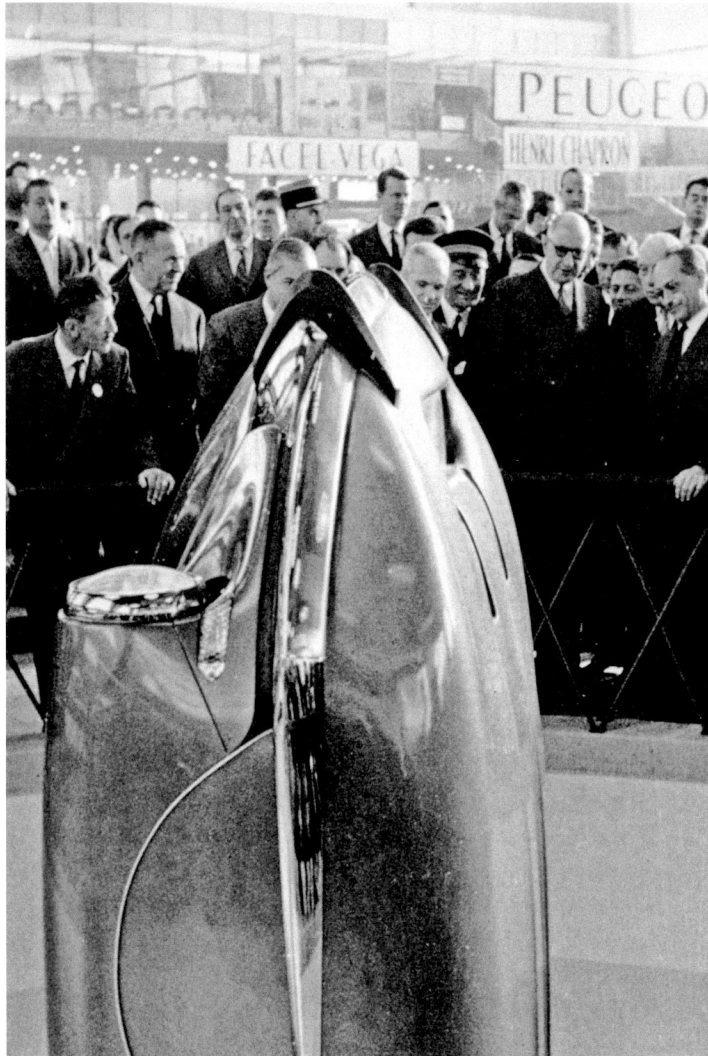

2. Semantic Vacuum?

The dissipation, the destruction of certain symbolic elements of the car, such as the radiator grille (which becomes a horizontal slit), the bumper (which becomes a bracket or seam between above and below) and the classic round headlight (which is replaced by an amorphous, elongated form) creates a "semantic vacuum" or "semantic openness." The round nose recalls a ship, the integration of the car's underside (front apron) turns the car into a fish, a ship, an airplane, while the roof-level blinkers become rockets, etc.

Fish, Ship, Airplane, Rocket

The car, or more precisely its form, retreats from a singular meaning. Bertoni said he was thinking of a fish while designing the DS. The car can be read in multiple registers. In its various presentations and in advertisements, the car is staged differently. In 1957 at the Milan Triennale, it was shown as a plane; the 1958 catalog depicted the Citroën as a ship; and in 1962, at the Paris Motor Show, it was a vertical rocket. Meanwhile, the body, underside and wheel fenders were all welded.

Paris Motor Show (1962)
Guest: President de Gaulle.

· Robert Dumoulin: Advertising (1958)
· Robert Dumoulin: Advertising (1957)

"Floating", air bed, inflatable raft

There is the special suspension, the hydropneumatic hydraulics, developed by Paul Magès. Each wheel is separately suspended, then conjoined via an oil-based ball bearing system partially filled with nitrogen and separated by a membrane. Using pressure, the volume of the carbon monoxide can be changed, thereby assuming the behavior of a mechanical spring. The car lies, or floats in a certain sense, on a bed of air, illustrated in advertisements in a variety of manners: water (oil) and air (nitrogen), ocean and sky, floating, carried by four rubber balls, a single person on an inflatable raft in the water, etc.

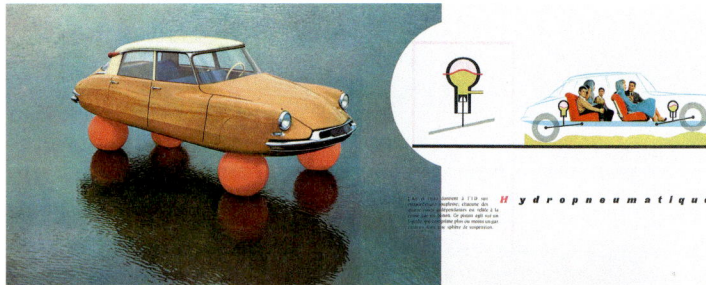

At the same time, the car can assume different heights or, expressed more precisely, postures. To accomplish this lift, a small amount of oil is pumped into circulation. When the car stops, the oil is pumped back into the reservoir and the car returns to its lowest position, a kind of resting state in which it "sleeps." Upon restarting, the car first has to get "into position." The car "lifts off," a quintessential phenomenon of the DS which has fascinated more than children (not unlike the inflation of an air mattress) as Jacques Tati's films prove. To change a tire, the car is put into its highest position, jacked up on one side and then reset in its lowest setting so that the wheels "retract" like those of an airplane.

See chapter: Advertising, pp. 176–179

Hydraulics

The hydraulic suspension was already imagined in the revised 2CV but was not implemented. The transmission and the integrated suspension that connected the front and rear wheels with a single spiral spring created the effect that, when the front wheel was compressed, the rear wheel expanded.

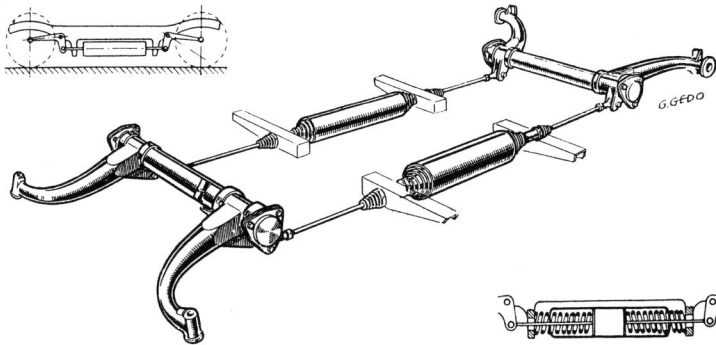

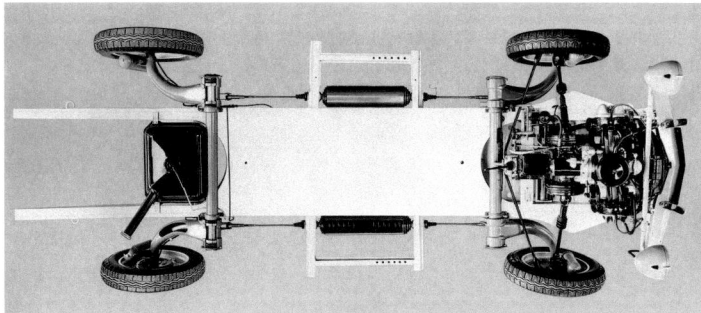

Citroën 2CV: Mechanical suspension

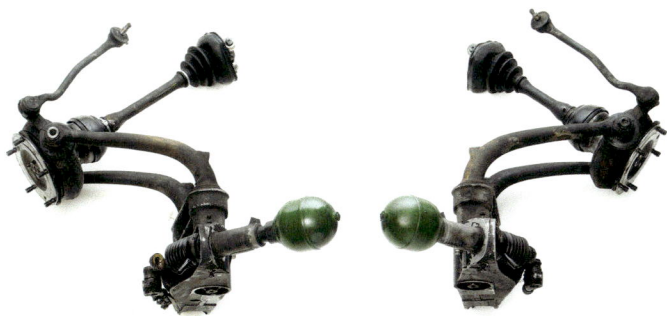

See chapter: Disassembly, pp. 108–109

hydraulique

DS 19

The hydraulic transmission rests upon a sphere filled with nitrogen (the car rests to a certain extent on a bed of nitrogen); the sphere and the pivot arms are connected via an oil circulation system (red). A second circulation system (yellow) within the ball regulates the height (low, medium, high and very high for a tire change).

Additional circulation systems are used for power steering, to support the brakes and the steering system and the switching.

konstante bodenfreiheit

veränderliche bodenfreiheit

Jede Änderung der Höhe zwischen Karosserie und Boden (Änderung der Belastung) setzt den automatischen Höhenkorrektor in Tätigkeit: Die Verbindungsflüssigkeit zwischen Kolben und Gas wird vermehrt oder vermindert die Bodenfreiheit ist wieder normal (16 cm).

Mit Hilfe eines einfachen Hebels kann der Fahrer die Bodenfreiheit nach Belieben bis auf 28 cm erhöhen, wenn ihm dies zum Überqueren einer sehr unebenen Strecke oder einer Furt notwendig erscheint.

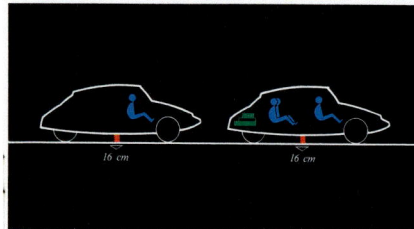

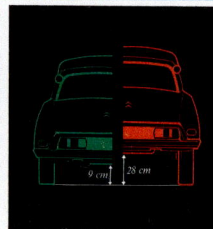

· Citroën DS with only three wheels
· See chapter: Advertising, pp. 182–183

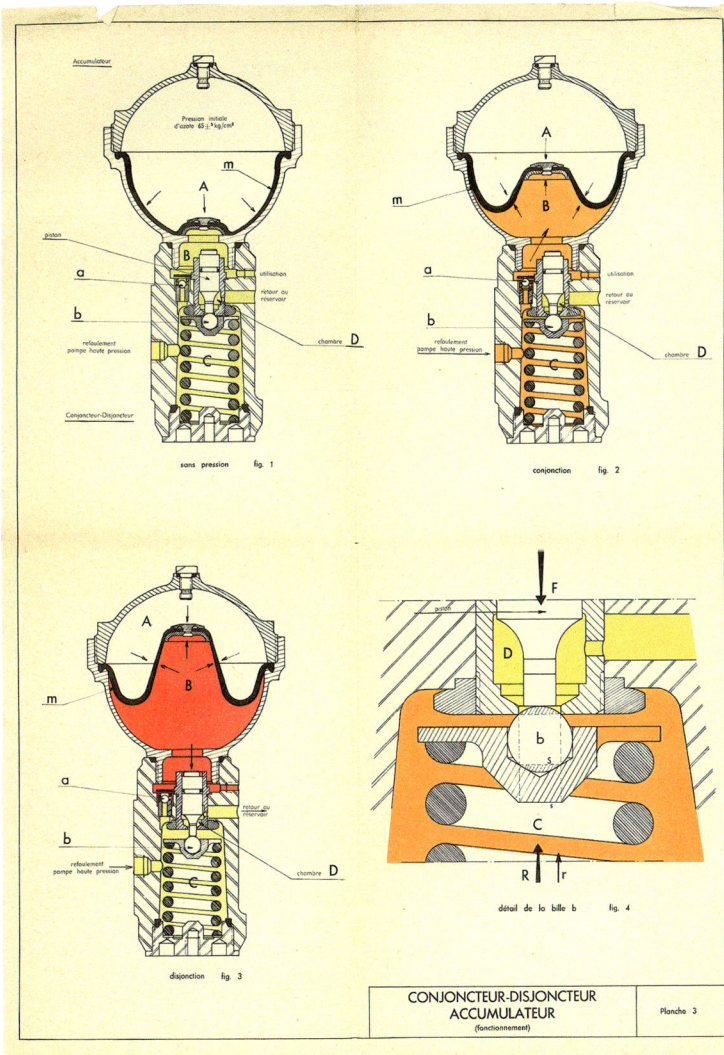

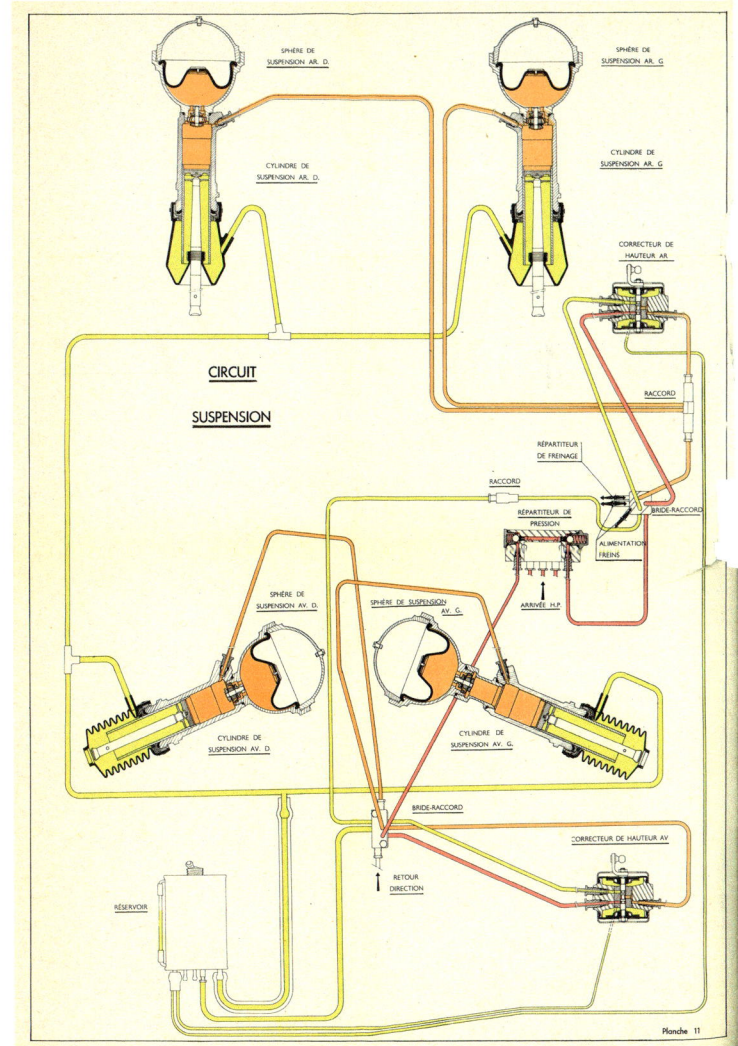

Left diagram (Planche 3):

Accumulateur

Pression initiale d'azote 65 ± ² kg/cm²

piston

a

b

refoulement pompe haute pression

chambre D

Conjoncteur-Disjoncteur

utilisation

retour au réservoir

m

A

B

sans pression   fig. 1

conjonction   fig. 2

disjonction   fig. 3

détail de la bille b   fig. 4

F

D

b

s

C

R   r

CONJONCTEUR-DISJONCTEUR ACCUMULATEUR (fonctionnement)   Planche 3

Right diagram (Planche 11):

SPHÈRE DE SUSPENSION AR. D.

SPHÈRE DE SUSPENSION AR. G

CYLINDRE DE SUSPENSION AR. D.

CYLINDRE DE SUSPENSION AR. G

CORRECTEUR DE HAUTEUR AR

RACCORD

CIRCUIT

SUSPENSION

RÉPARTITEUR DE FREINAGE

RACCORD

RÉPARTITEUR DE PRESSION

BRIDE-RACCORD

ALIMENTATION FREINS

ARRIVÉE H.P.

SPHÈRE DE SUSPENSION AV. D.

SPHÈRE DE SUSPENSION AV. G

CYLINDRE DE SUSPENSION AV. D

CYLINDRE DE SUSPENSION AV. G

BRIDE-RACCORD

RETOUR DIRECTION

CORRECTEUR DE HAUTEUR AV

RÉSERVOIR

Planche 11

Hydraulic transmission

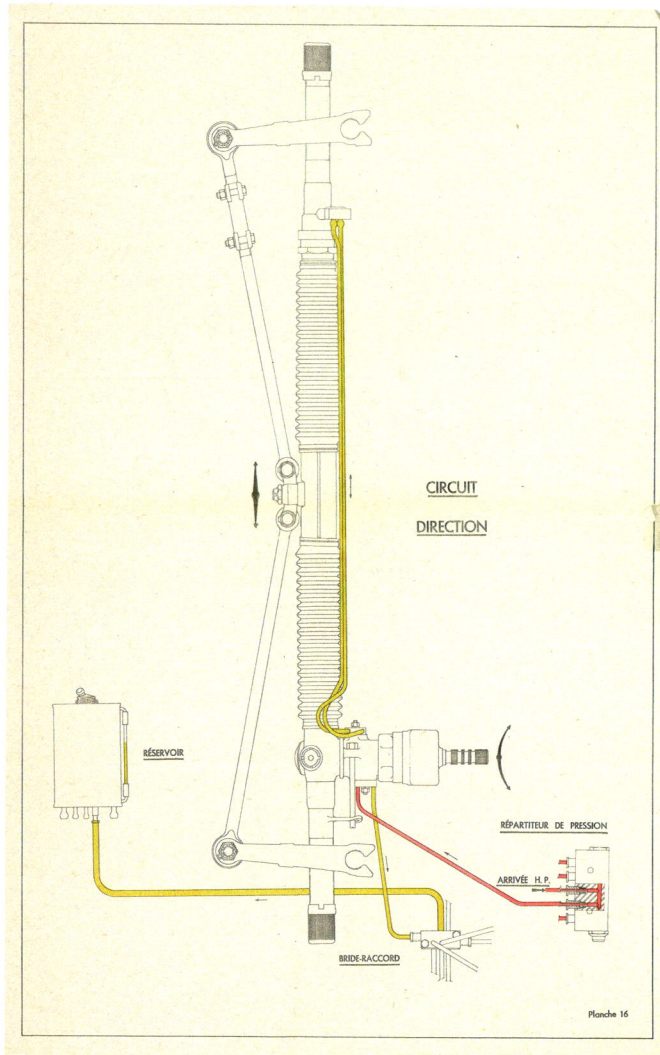

CIRCUIT

DIRECTION

RÉSERVOIR

RÉPARTITEUR DE PRESSION

ARRIVÉE H.P.

BRIDE-RACCORD

Planche 16

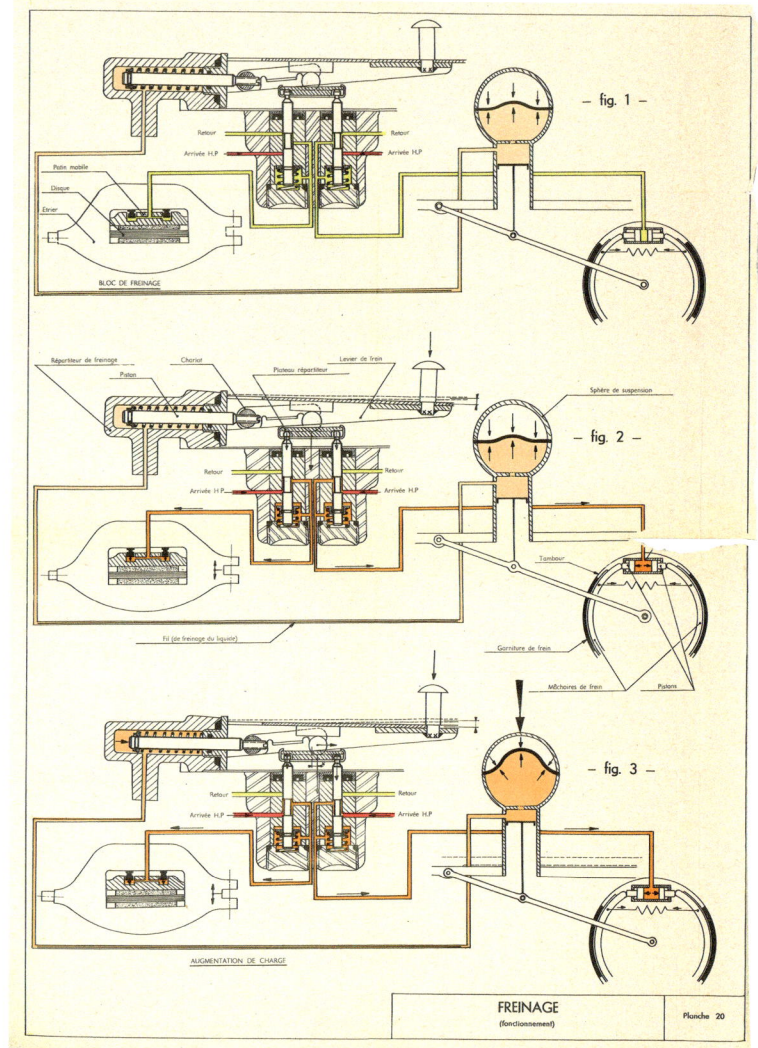

- fig. 1 -

Retour   Retour
Arrivée H.P   Arrivée H.P
Patin mobile
Disque
Étrier

BLOC DE FREINAGE

Répartiteur de freinage   Chariot   Plateau répartiteur   Levier de frein
Piston   Sphère de suspension
Retour   Retour
Arrivée H.P   Arrivée H.P
Tambour
Fil (de freinage du liquide)   Garniture de frein
Mâchoires de frein   Pistons

- fig. 2 -

Retour   Retour
Arrivée H.P   Arrivée H.P

- fig. 3 -

AUGMENTATION DE CHARGE

FREINAGE
(fonctionnement)

Planche 20

Steering   Brakes

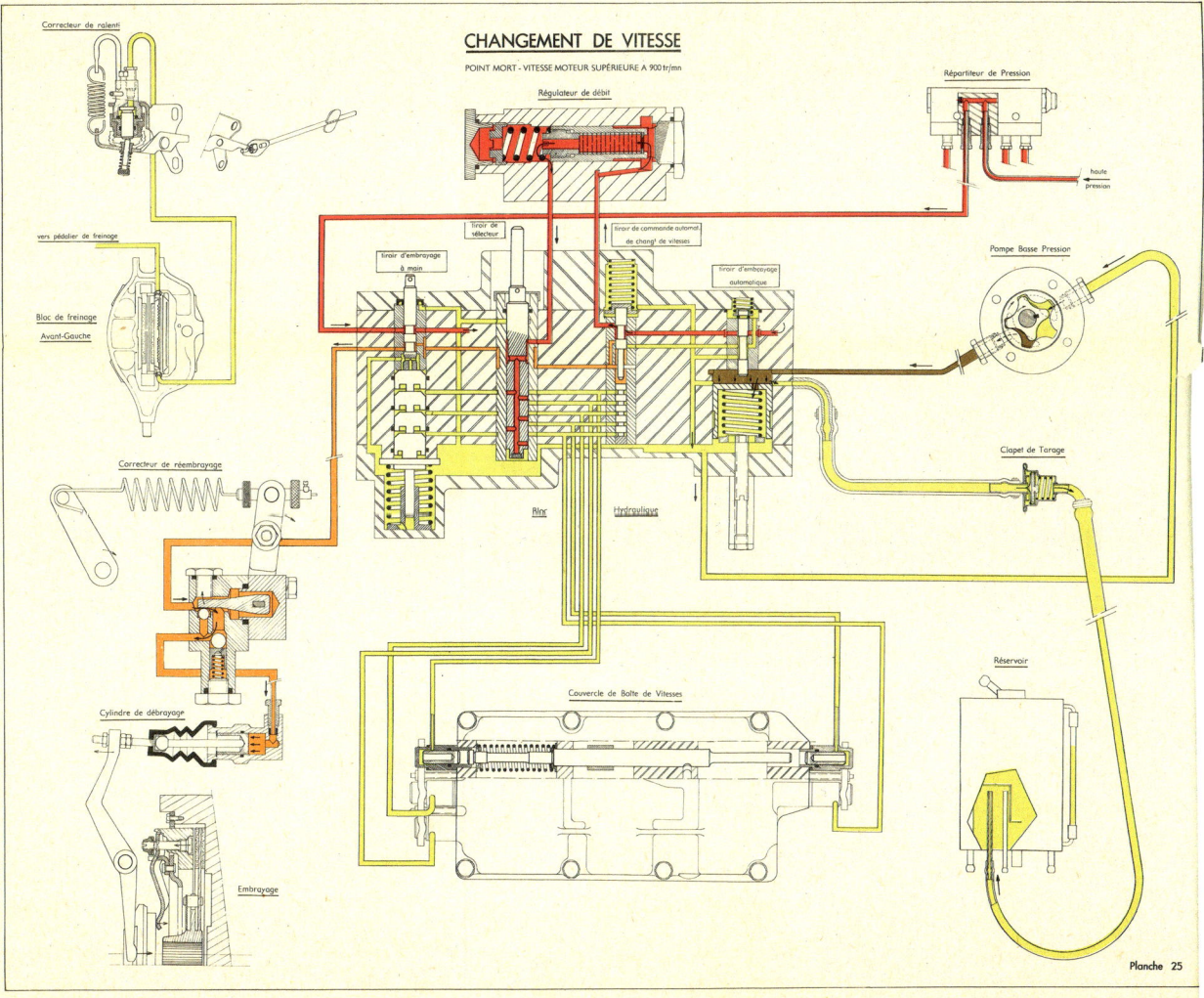

CHANGEMENT DE VITESSE

POINT MORT - VITESSE MOTEUR SUPÉRIEURE A 900 tr/mn

Correcteur de ralenti

Répartiteur de Pression

haute pression

Régulateur de débit

vers pédalier de freinage

Bloc de freinage Avant-Gauche

tiroir de sélecteur

tiroir de commande automat. de chang! de vitesses

tiroir d'embrayage à main

tiroir d'embrayage automatique

Pompe Basse Pression

Correcteur de réembrayage

Clapet de Tarage

Bloc   Hydraulique

Cylindre de débrayage

Réservoir

Couvercle de Boîte de Vitesses

Embrayage

Planche 25

Switching

150

Double arrow and motion

In advertisements, the Citroën logo of a double arrow that derives from a wedge-shaped cog[1] is reinterpreted as a double arrow that clears a path through the air. Roland Barthes wrote about this: *The emblem with its arrows has in fact become a winged emblem, as if one were proceeding from the category of propulsion to that of spontaneous motion, from that of the engine to that of the organism.*[2]

The advertisement also uses suggestively archaic arrow-like artifacts to recall Bertoni's sculptures.

1 – In 1900, André Citroën assumed the patent to produce this kind of cog.
2 – Barthes, op. cit.

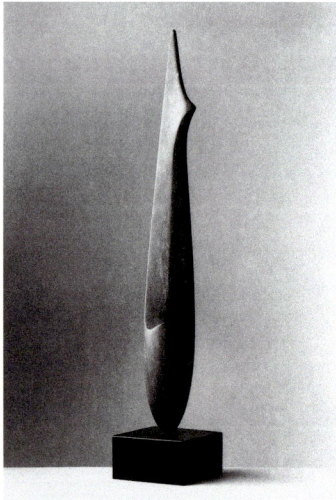

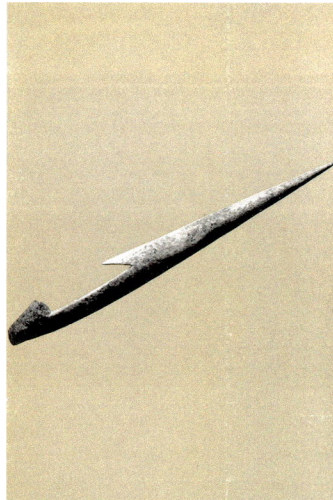

· *Flaminio Bertoni: Verticale (1952)*
· *Advertising (1958)*

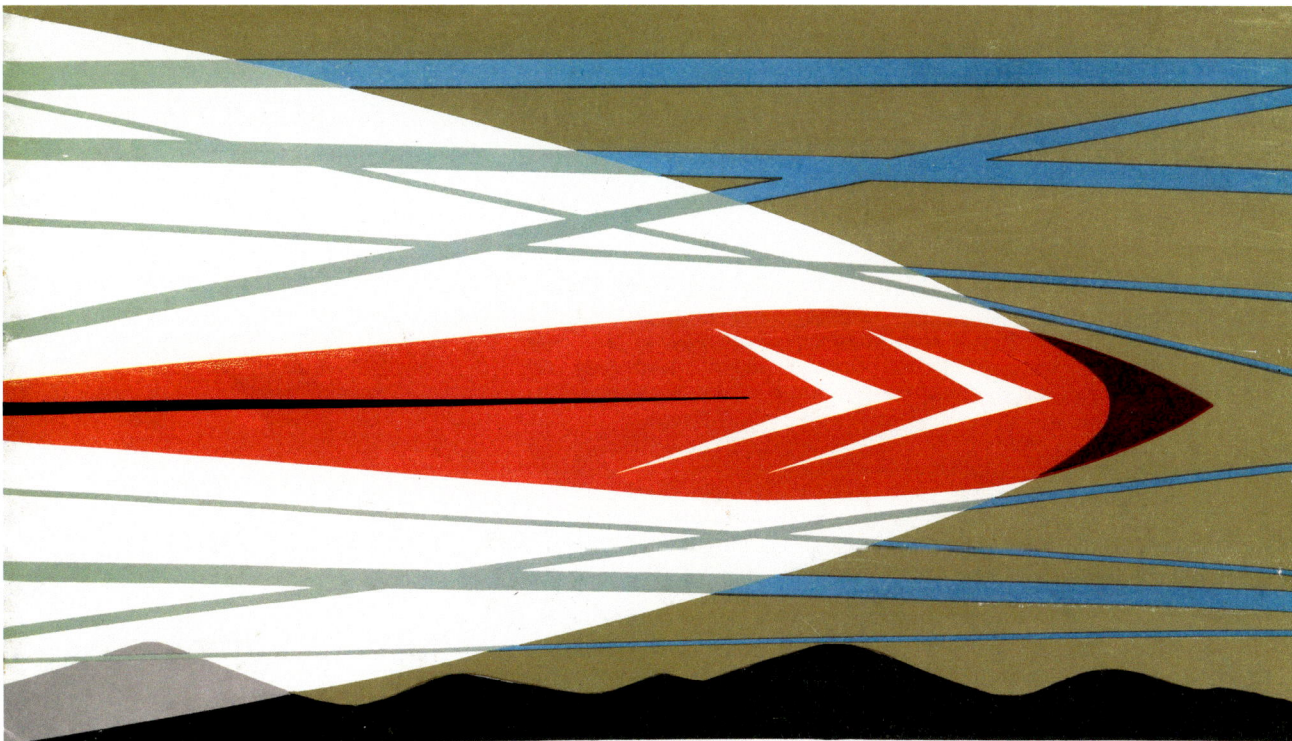

Hydropneumatique **DS 19**
CITROËN

Advertising (1957)

3. Max Bill
Beauty from Function and as Function

This phrase was the title of a 1948 lecture delivered to the Swiss *Werkbund*. Bill conditionalized and expanded upon the classical Functionalist formulation "Form Follows Function" in two respects. First, rather than juxtapose function to form directly, he juxtaposes it to beauty. Secondly, he also postulates "beauty" as the central function of form "in itself" (of course, given the consideration and fulfillment of functional criteria).[1]

An illustration of this concept is his irradiation lamp (1951). The lampshade and its base, which houses its transformer, have the same form despite their different functions. In the process, the "design-determining function" of form is valued above effective function, by contrast to the classic table lamp of the Swiss manufacturer BAG, for example. The flexible connective conduit, which functions as a handle for carrying the lamp, connects the two identical forms and their different functions into a dumbbell-shaped object.

Form, Function, Beauty = Gestalt [2]

In 1956, Bill expanded on the concept and created a "three-pack" of "Form, Function, Beauty" against which he set "Design." The concepts were thus freed from their direct causality and allowed to generate an open operational "field of form-finding."

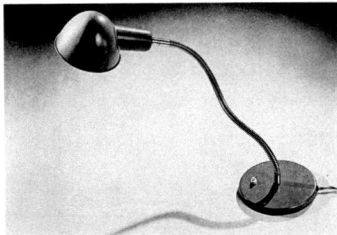

Classic table lamp by BAG (Bronzewarenfabrik AG)

1 – See Stanislaus von Moos, "Recycling Max Bill," in *minimal tradition*, op. cit. The text is also published in: Lars Müller (ed.), *Max Bill's View of Things. Die gute Form: An Exhibition 1949*, Lars Müller Publishers (Zürich 2015), p. 143 ff. Claude Lichtenstein cross-references this idea with Vitruvius' three precepts: *Firmitas* (solidity), *Utilitas* (utility) and *Venustas* (delight).
2 – Max Bill, "(Form, Funktion, Schönheit) = (Gestalt)" in: *Max Bill*, exhibition catalog, Ulm 1956.

Max Bill: Irradiation lamp "Höhensonne" (1951)

4. Flaminio Bertoni

Flaminio Bertoni's design work corresponds to the fundamental concepts sketched by Bill, as demonstrated in the preceding analysis. The form-finding process "frees" itself from technical and functional stipulations and assumes, in the sense that Max Bill intended, a certain degree of "autonomous status." The design overcomes (or elevates) technical demands and lends the automobile its independent character. This is true for all four of Bertoni's cars. The solution is always more than the representation of what is inside. "Form, Function, Beauty = Gestalt," all four assume an equilibrated relationship to one another while respecting the technical demands of the work, which are in this case:

- **Motor is located in front of the drive axel:** This results in a long, somewhat awkward nose in the VGD, which is abandoned in the DS. By locating the reserve tire above the front axel as a buffer, Bertoni, in collaboration with the engineers, "rescued" the elongated, rounded ship-like nose. At the same time, the form became more compressed and, in the process, more elegant.
- **Comfortable boarding** translated into a rear wheel set further back: by means of an additional facet in the side view and the housing on the rear wheel, Bertoni stretches the DS's silhouette and makes it even more dynamic.
- **Increased track in front and lesser track in back for improved driving quality, the tractix:** Bertoni uses the tapering of the overall dimension from front to rear in order to coordinate seams and lines.
- **Better headroom for the rear seats:** Bertoni "decimates" the dogma of the descending streamline and creates an all-encompassing 360-degree panorama window with completely new inside/outside relationships. Roof and curved window become a cockpit.
- **Improved road illumination at night:** In other words, movable high beams and low beams. The Citroën team used this concept "to overcome" the classic, point-based light source. Like the radiator grille, the headlights disappear from the car's front and become part of the car body. The moveable headlights which light the road perfectly and follow the road's trajectory, would in 1967 become the car's second technical trademark, along with its hydraulic suspension. The headlights become eyes, and the mysterious cat's eyes emerge.

Flaminio Bertoni
(1903 – 1964)

In an interview in 1959[1], Bertoni discussed the balancing act required to "reconcile" technology and design. He talks about the priority of form in conjunction with beauty as a function, and about his efforts to defend these priorities against all the technical demands.

1 – Published in: Leonardo Bertoni, op. cit., p. 114:

F. Posso chiederle, caro sig. Bertoni, come sia giunto alla carrozzeria?

B. Sono nato carrozziere! Per me il bisogno di disegnare un'autovettura è lo stesso che prova un pittore di dipingere o un musicista di comporre una melodia. Personalmente le mie carrozzerie le ricerco nella natura. Trent'anni fa ne ricavai una da un cigno. La DS mi è stata ispirata da un pesce.

F. Preferisce scolpire o disegnare carrozzerie?

B. Non c'è differenza tra le due cose, tutto ciò che è volume è scultura, la carrozzeria ha un volume: Le due cose sono identiche.

F. Perciò quella del carrozziere è un'arte?

B. Qualche cosa di più, in quanto c'è la creazione, oltre alle imposizioni estranee, come le esigenze meccaniche che non hanno nulla a che vedere con l'arte.

F. Come nasce una carrozzeria?

B. La carrozzeria è un nucleo, un insieme che si forma nella mente del carrozziere: poi, purtroppo, si comincia a rovinare quel nucleo meraviglioso per aggiungervi le ruote, i vari aggeggi, quella orribile cosa che è il motore.

F. Perché il motore è una cosa orribile?

B. Beh, in un certo senso rovina tutto, perché rappresenta la tecnica, e la tecnica è implacabile: alle sue esigenze si deve sottomettere anche l'ispirazione.

F. Ci vuole molto tempo per creare un prototipo?

B. A volte anche dieci anni, uno ha un'idea, poi l'abbandona, poi la riprende… proprio come un quadro o un romanzo. Poi fatto il prototipo, cominciano i guai, perché bisogna adattarlo alle esigenze della fabbricazione di serie, che a volte fanno a pugni con l'idea originaria del carrozziere.

F. Ma dev'essere una gioia progettare una carrozzeria!

B. Si, quanto si tratta di fuoriserie. Allora l'estro e la linea possono sbizzarrirsi, la fantasia può vagabondare fuori di espressioni tecnico-commerciali.

F. Can I ask you, dear Mr. Bertoni, how did it get to the bodywork?

B. I was born a coachbuilder! For me to design a car is the same as for a painter or a musician to compose a melody. Personally, for my bodywork I look in nature. Thirty years ago I got an idea from a swan. The DS was inspired by a fish.

F. Do you prefer to sculpt or design bodywork?

B. There is no difference between the two, everything that is volume, is also sculpture, the body has a volume: the two things are identical.

F. Why is the bodywork of a car an art?

B. It is something more. There may be a concept, in addition to the external impositions, like the mechanical needs that have nothing to do with art.

F. How is a bodywork born?

B. The bodywork is a concept, a whole that is formed in the mind of the body: then, unfortunately, it begins to ruin that wonderful concept to add wheels, various gadgets, that horrible thing that is the engine.

F. Why is the engine a horrible thing?

B. Well, in a certain sense it ruins everything, because it represents technology and technology is relentless: inspiration must also be subdued to its needs.

F. Does it take a long time to create a prototype?

B. Sometimes even ten years, one has an idea, then abandons it, then takes it back... just like a painting or a novel. Once the prototype was made, the trouble started, because we had to adapt it to the needs of mass production, which sometimes clashed with the original idea of the coachbuilder.

F. But it must be a joy to design a body!

B. Yes, when it comes to custom-built cars. Then the inspiration and the line can indulge, the fantasy can wander out of technical-commercial expressions.

5. Otl Aicher

An important critique of the Citroën DS came almost thirty years after Roland Barthes' text. Its author was Otl Aicher, a professor beginning in 1955 at the Ulm Hochschule für Gestaltung founded by Max Bill. His book is entitled *Kritik am Auto* (Critique of a Car). Following an exhaustive analysis, he concludes, *and nonetheless, this car is a milestone in an incorrect development. ...* His harshest critique targets the interior space *as a volumetric remainder of the streamlined body* while also making a point about the CW's "poor" aerodynamic value of 0.38, a value higher than the 0.30 measured before the war on a similar car in comparable testing.[1]

In 1984, the year that Aicher's book was published, there were signs of a paradigm shift in car design, particularly with Giorgetto Giugiaro's VW Golf (1974) and Fiat Uno (1981). "Narrow and high" was the new leitmotif, in contrast to the long-standing "wide and low" maxim that had guided streamline form.[2] Carefully Aicher describes this development.

1 – Claude Lichtenstein offers a comprehensive overview of streamline technology in: *Stromlinienform*, op. cit.

2 – Aicher's brief sketch of the car's (incorrect) development story represented by streamline design is summarized below for its central arguments: In 1921, Paul Jaray defined the optimal basic form of streamline design as the contour of a wing he had first used in his automobile (1922): a compartment, wing-shaped in plan, "sits" on a tail whose sectional form is wing-shaped. The classic radiator grille has also been reduced to purely technical air intake. In essence, streamline design is the concept of the pontoon-shaped body which combines the various parts of the car – passenger compartment and tail as well as head and taillights and grille – in an overall form. In the early 1930s, studies by, among others, Kamm and Everling in Germany and Lay in the US indicate that a cropped teardrop shape, truncated at the back, performs better in a wind tunnel than a full teardrop form. This is an important insight that is nonetheless ignored by the "stylists," as Aicher condescendingly calls them. The maxim of the years that followed still remained "low and elongated." It is only in the early 1970s, with the "Hard Edge" car body and its characteristic tail and "line breaks" that the insights of the 1930s are picked up again and further developed. Aicher calls Giugiaro's Fiat Uno (1981) with its steep, stepped tail the high point of this development. Because of the increasing use of laterally positioned motors, the front portions of the new cars (hood) become even shorter.

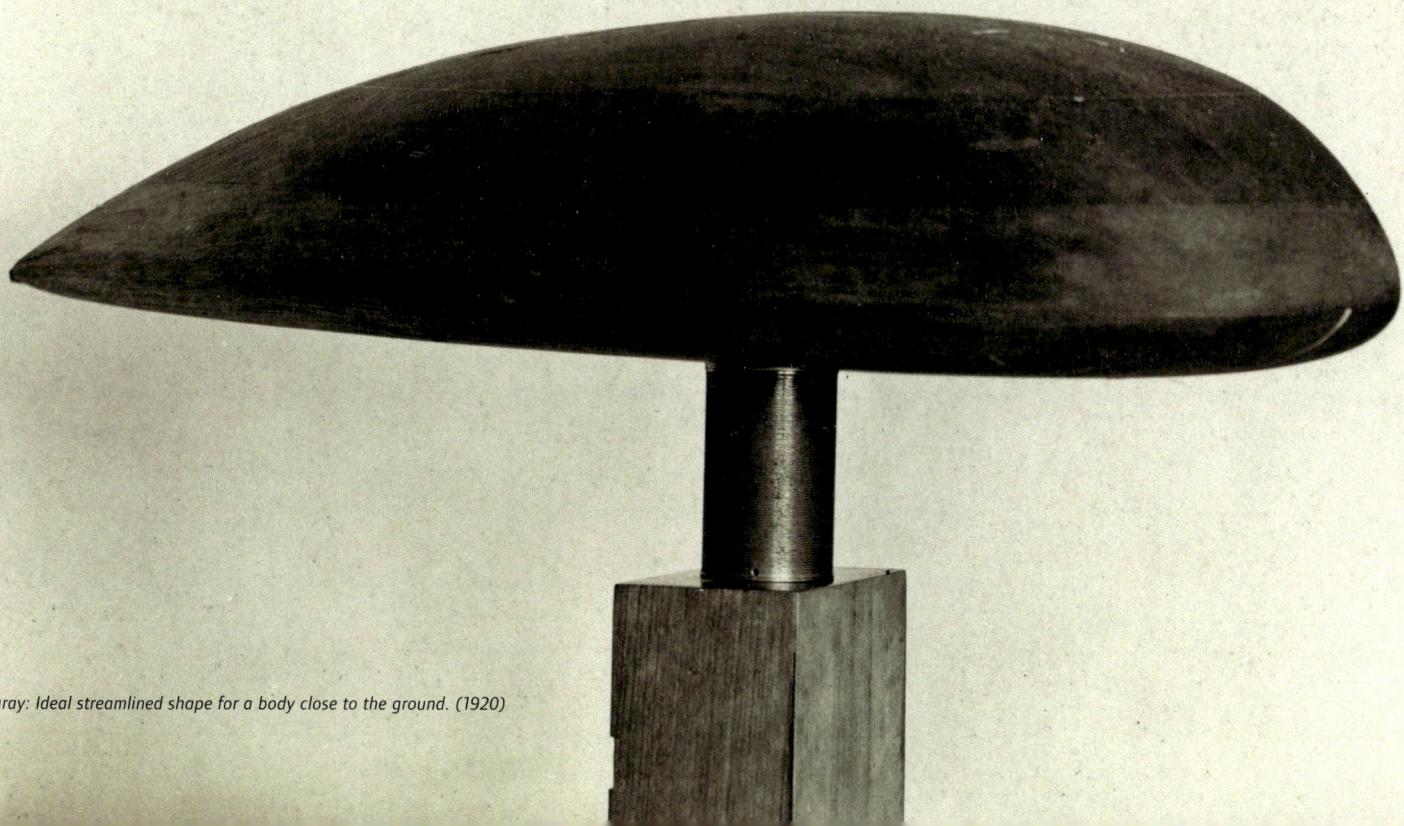

Paul Jaray: Ideal streamlined shape for a body close to the ground. (1920)

160

das ergebnis von jarays wissenschaftlichen untersuchungen, die durchdringung zweier strömungsformen, einer für fahrwerk und motor, einer als fahrgastkabine, wird in prototypen bzw. einer patentschrift aus dem jahr 1922 vorgestellt. links unten ein analog konstruierter mercedes benz rekordwagen von 1934.

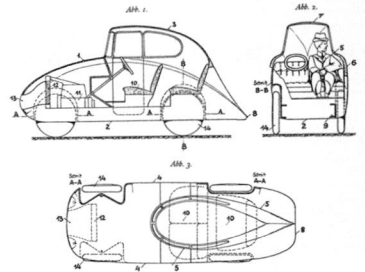

12

Extract from: Otl Aicher, "Kritik am Auto," pp. 12–13

te er eine gondelartige kabine. der nutzen bestand
hst nicht so sehr in der erhöhung der geschwindigkeit,
em in der verhinderung von staubwirbeln auf den schot-
aßen. 1921 meldete er ein voll verkleidetes stromlinien-
rum patent an. bereits ein jahr später machte er versu-
uf der avus in berlin und erreichte bei 20 PS eine
windigkeit von 100 km/h, ein jahr später sogar 130 km/h.
folg war eindeutig, was die technische seite anbelangte.
rrringerung des luftwiderstandes spart energiekosten,
glicht höhere geschwindigkeiten und stabilisiert das
aber die stromlinienform ging zu lasten der kabine. sie
e immer mehr reduziert und beschnitten, vor allem, nach-
man das konzept langgezogener autos aufgegeben hatte
ürzere formen suchte.
as dilemma des automobilbaus zeigte sich in aller deut-
it. auf der einen seite sollte das auto ein optimaler
lter zum transport von bequem sitzenden personen sein.
er anderen seite zwang eine günstige stromlinienform
e zur reduktion der kabine. sie sollte flach und nach
n abfallend sein.
nlich wie buckminster fuller versuchte karl schlör 1937
esamtkörper des wagens so groß zu machen, daß die
setzte kabine verschwand und die passagiere in einer
n tropfenform saßen. aber das auto wurde zu voluminös.
ndessen entstanden in den frühen dreißiger jahren zahl-
versuchsanstalten und büros, meist an technische
chulen angegliedert, die windkanalstudien betrieben, so
n berlin unter der leitung von prof. emil august everling
ne in stuttgart unter leitung von prof. wunibald kamm,

im gegensatz zu jarays an-
satz wurde vor allem in den
USA an dem versuch festge-
halten, das auto als eine ein-
heitliche strömungsform auf-
zufassen. links ein schlör-
wagen von 1938, unten
buckminster fullers dyma-
xion car von 1004.

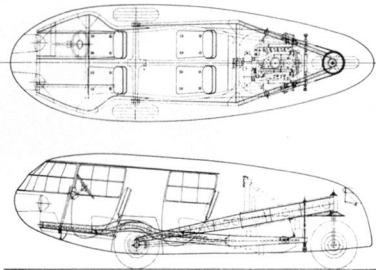

13

162

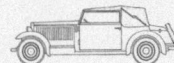

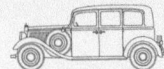

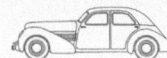

e. neumann, 1910
adler, gropius, 1930
mercedes 260 D, 1936
citroën 7 traction, 1934
cord 810 beverly sedan, 1936
chevrolet master deluxe, 1939

tatra 87, 1936
BMW kamm, 1938
studebaker deluxe starlight, 1951
mercury park lane, 1959
citroën DS 19, 1956
citroën CX 25, 1974
NSU RO 80, 1967
mercedes 190, 1983

92

Extract from: Otl Aicher, "Kritik am Auto," pp. 92–93
The cars of Flaminio Bertoni and Giorgetto Giugiaro are highlighted (gray)

aktuellste beitrag zur entwicklung des automobils ist
e das sogenannte hochdachauto. bis jetzt am eindeutig-
realisiert im FIAT uno. es vereinigt drei qualitäten:
nstiger fahrgastraum für aufrechtes sitzen, kopffreiheit
d bequemes ein- und aussteigen,
te motorisierung mit geschwindigkeiten weit über den
mpolimits,
nstige strömungsform, besser als bei herkömmlichen
ortwagen, trotzdem stark reduzierte länge.
ist man damit nicht zum ausgangspunkt der automobil-
wicklung zurückgekehrt, als man noch keine stromlinien-
kannte? der FIAT uno hat viele gemeinsamkeiten mit
m auto von e. neumann aus dem jahre 1910. war der
des automobils nicht eine bewegung im kreis, bei der
beim ausgangspunkt, wenn auch auf einem anderen
au, ankam?
zunächst entwickelte sich das auto von der handwerk-
gefertigten kutsche zum industrieprodukt. der karosse-
engler wurde durch schwere blechpressen ersetzt, die
eformte teile in serie ausstießen.
das stilistische leitbild des automobilbaues wurde für
albes jahrhundert die tropfenform, die mehr und mehr
der keilform abgelöst wird.
auf dem gebiet der mittelklasse- und kleinklassewagen
og sich die entwicklung zu einem neuen konzept. quer-

gestellter motor und verzicht auf langen kofferraum führen
zu kürzeren, handlicheren autos. das auto bekommt eine
hecktür für bequemes laden. die rückbank kann alternativ
als ladefläche benutzt werden.
   wichtige anregungen kommen von vierradangetriebe-
nen allzweckautos. der einfluß geht in richtung auto als
gebrauchsgegenstand statt als demonstrationsobjekt.

käfer, 1939
ën 2 CV, 1949
  500, 1957
ris mini, 1959
golf, 1974
uno, 1983

jeep, 1941
land rover, 1948
FIAT panda, 1980
range rover, 1970
renault espace, 1984

93

6. Gio Ponti

In 1953, at the same time that Bertoni was developing the design of the DS, Gio Ponti was also developing a car: the "Diamond" and its "Linea Diamante."[1] The first versions referenced the drop-shaped streamline form. Soon, he moved away from this approach and noted: *It was the era of inflated forms, the doors were quite thick. ... the doors host space where we don't need it and the machines become more cumbersome* and he demands:

we have to find a line… He describes his design as follows: *…with its diamond surfaces this car has thin doors, and they are… vertical and glazed, very big, can get down completely – this car is very light with clear color inside, a lot of light, a lot of air, a lot of interior space, the roof is high.*

The shift from the soft line of streamline design to a more faceted contour (Diamant) is an impressive anticipation of the "Hard Edge Design" of the "high-roofed cars" Aicher described. The design of the DS (1955) by Flaminio Bertoni coincides with the high point of streamline design but is simultaneously the prelude to its demise. Gio Ponti was, incidentally, an enthusiastic Citroën DS driver.

1 – A model of the car was reconstructed for the exhibition Grand Basel in 2018. The text fragments are extracted from notes on the drawings in the exhibition: "C'était l'époque des formes gonflées, les portes avaient une grande épaisseur. / Les autos sont dans la période la ligne gonfle, les portières hôtent l'espace où l'on en a (pas) besoin et les machines deviennent plus encombrantes: avec leurs "vides gonfles" dans les portières et au dessus des r(o)ues sont des machines en crinoline." "On doit arriver à une ligne / avec ses surfaces en diamant cette voiture a des portières minces, et elles sont / verticales et la surface vitrée, très grande peut descendre complêtement – cette voiture était de couleur très clair à l'intérieur; beaucoup de lumière, beaucoup d'air, beaucoup d'espace intérieur, le toit était haut." The translation is shortened. See also Lisa Licitra Ponti, *Gio Ponti. The Complete Work 1923–78*, MIT Press, Boston 1990, p. 167.

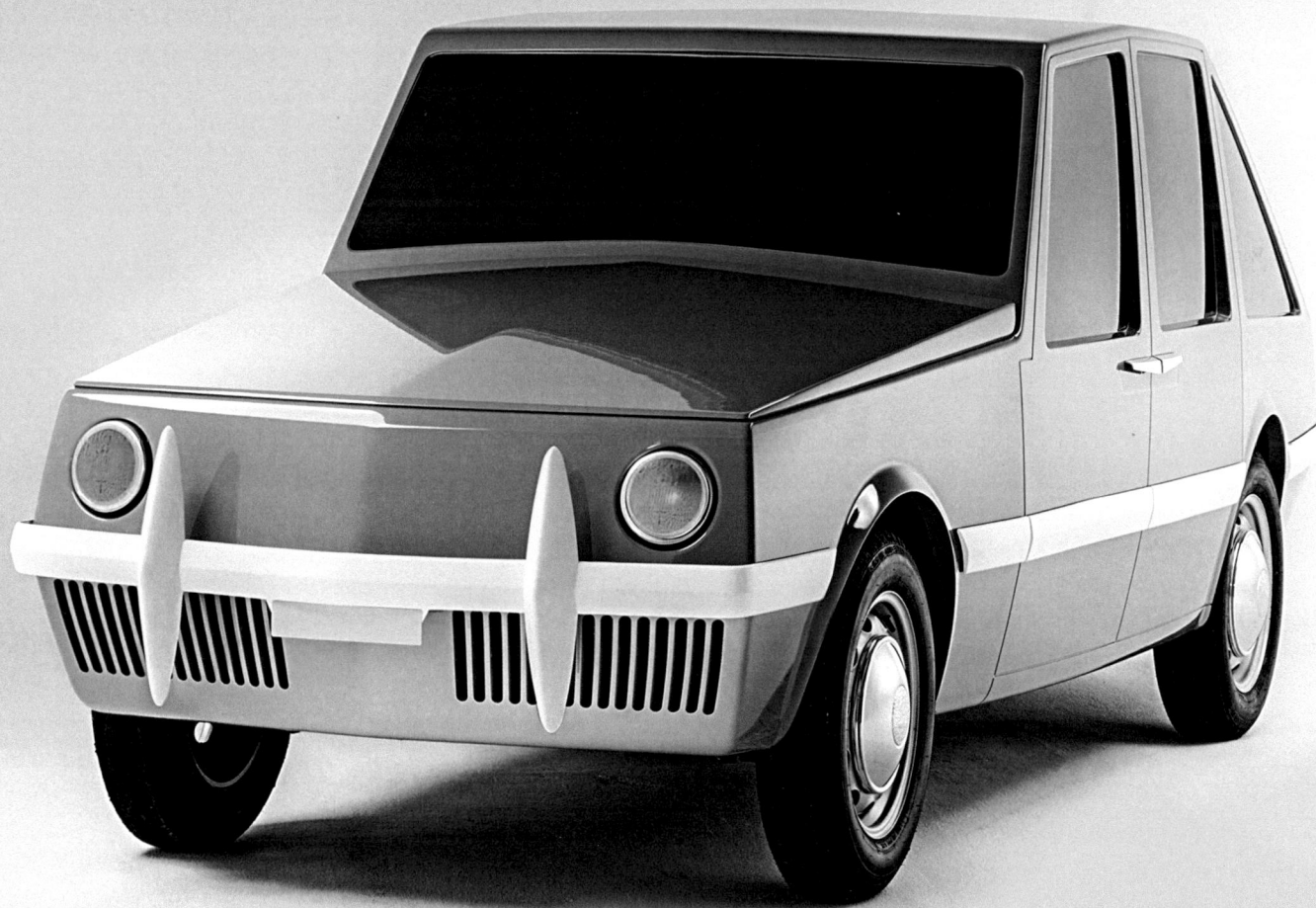

Gio Ponti: Diamond (1953)

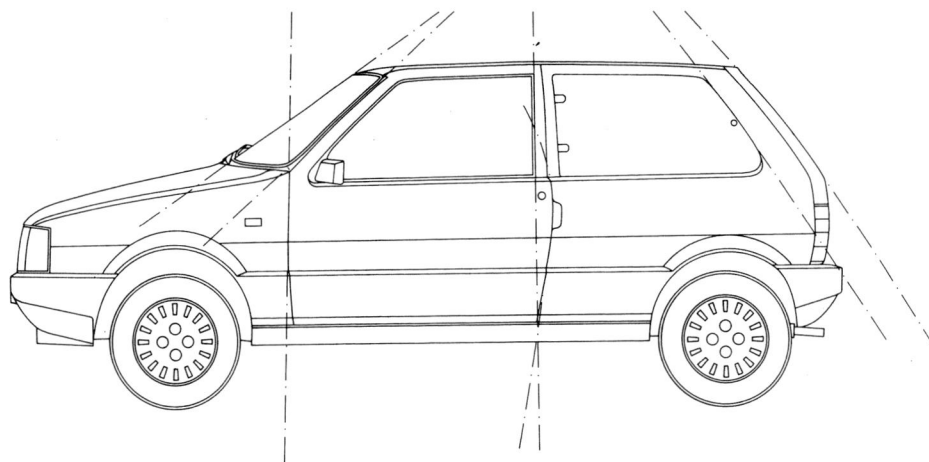

Extract from: Otl Aicher, "Kritik am Auto"
Giorgetto Giugiaro: VW Golf (1974) and Fiat Uno (1981)

Gio Ponti: Diamond (1953)

Advertising

Advertising

The marketing, the entrée, the launch of the 1955 DS is a publicity stroke of genius, a milestone in automobile advertising. The best advertising agencies, like Robert Dumoulin and Robert Delpire, designed the advertising brochures and thematized the car's innovations: streamline (good aerodynamics), hydraulic suspension / comfort, panoramic view (2.25 m² of glass), safety, different color combinations for the car body and interior. In 1967, the advertising becomes sportier, younger and more aggressive – like Helmut Newton's photos – and the palette of models more differentiated.

Helmut Newton (ca. 1969)

FICHE TECHNIQUE
DE L'ID (LUXE ET CONFORT)

MOTEUR - 4 Cylindres 78 x 100 mm - Cylindrée : 1911 cm³ - Compression : 7,5
Puissances : effective 66 ch ; administrative 11 CV - Couple maximum : 13,5 m/k - Graissage : sous
pression par pompe mécanique - Carburateur : Solex 34 - Tubulures intégrées - Alimentation du carburateur
par pompe mécanique - Batterie : 6 V 75 A/h - Démarreur électrique commandé par relai - Dynamo 180 W à
régulateur - Refroidissement par pompe et thermostat - Ventilateur nylon.

BOITE DE VITESSES - 4 vitesses (2°, 3°, 4° synchronisées) + AR - Levier de commande sous le volant.

DIRECTION à crémaillère - Volant monobranche garni d'un enroulement plastique.

TRANSMISSION - Traction avant - Joints BIBAX - Couple conique 9 x 35 à taille hélicoïdale - Embrayage monodisque à sec.

CHASSIS - Plateforme à longerons latéraux en tôle d'acier soudée - Plancher plat.

SUSPENSION - 4 roues indépendantes munies chacune d'un bloc hydropneumatique de suspension avec amortisseur intégré - Barres
anti-roulis - Correcteurs d'assiette.
ROUES : Fixation centrale.
PNEUS X : 165 x 400 à l'AV, 155 x 400 à l'AR. Roue de secours 155 x 400 utilisable temporairement à l'avant.

FREINS - FREIN PRINCIPAL : à disques sur roues AV, à tambours sur roues AR - Commande hydraulique par pédale - Compensateur
automatique de l'usure des garnitures AV.
FREIN SECONDAIRE : Mécanique à main gauche sur disques à l'AV - Verrouillage de sécurité à l'arrêt.

POIDS ET ENCOMBREMENT - Poids (avec 5 litres d'essence, outillage et roue de secours) : 1090 Kg - Empattement :
3,125 m - Voie AV 1,50 m - Voie AR 1,30 m - Longueur hors-tout : 4,80 m - Largeur hors-tout : 1,79 m - Hauteur hors-tout : 1,47 m - Rayon
de braquage : 5,50 m.

CAPACITÉ - Réservoir à essence : 65 litres - Carter moteur : 4 litres - Carter boîte : 2 litres (boîte et pont) - Volume de la malle :
0,500 m³ entièrement utilisable.

CONFORT - Chauffage et dégivrage par air pulsé, radiateur spécial, 2 bouches à l'AV - Ventilation par arrivée d'air à
orientation réglable à droite et à gauche de la planche de bord - Sièges avant réglables.

PERFORMANCES - Vitesses maxima : 40 Km/h en 1°, 80 Km/h en 2°, 115 Km/h en 3°, 135 Km/h en 4°
Consommation : 9 litres aux 100 Km à 75 Km/h de moyenne.

CARACTÉRISTIQUES PARTICULIÈRES DE L'ID CONFORT

Sellerie : Sièges couchettes en jersey de nylon ou hélanca. Accotoir central à l'arrière.
Aération et chauffage : à l'avant : deux orifices d'aération supplémentaires (au
plancher), à l'arrière : sur longeron droit un orifice de chauffage
supplémentaire. Double lave-glace

Société Anonyme André Citroën, 133, Quai André Citroën - R. C. Seine 54 B 9455 - N° d'Entreprise 261 75 115 0001 N...

L'embarras du choix : Le jeu des différents coloris de la caisse et des

de l'ID permet plus de 50 combinaisons

Achevé d'imprimer en Octobre 1959 sur les Presses de la Photolith L. Delaporte, Paris.

Création de Robert Dumoulin. Photographies de Pierre Jahan.

Robes de Jacques Esterel, réalisées dans les tissus de l'I. D. 19.

L'Air et l'Eau donnent à l'ID son
extraordinaire souplesse; chacune des
quatre roues indépendantes est reliée à la
caisse par un piston. Ce piston agit sur un
liquide qui comprime plus ou moins un gaz
contenu dans une sphère de suspension.

Hydropneumatique

2 , 2 5   m²   d e   g l a c e s ,   v o u s   ê t e s   l e   c o n d u c t e u r   e u r o

qui voit le mieux

ds

grosse elastizität

Jedes der unabhängigen Räder ist mit dem Fahrgestell über einen Schwingarm mit einem Kolben verbunden. Dieser Kolben wirkt innerhalb eines Zylinders auf eine Verbindungsflüssigkeit, die mehr oder weniger – je nach den Bewegungen des Kolbens – eine konstante Gasmenge in einer am Fahrgestell angebrachten Kugel komprimiert. Eine vollkommen dichte, elastische Membrane trennt die Gasmasse von der Flüssigkeit. Jede senkrechte Bewegung... des Rades betätigt also den Kolb[en] der die Verbindungsflüssigkeit in [den] Zylinder und den unteren Teil d[es] Federelementes drückt, wo sie d[as] Gas, je nach der Ausgangsbeweg[ung] des Rades, mehr oder weniger komprimiert. Das Gas reagiert, gemäss den Eigenschaften jeder pneumatischen Federung, mit ausserordentlicher Geschmeidigk[eit]

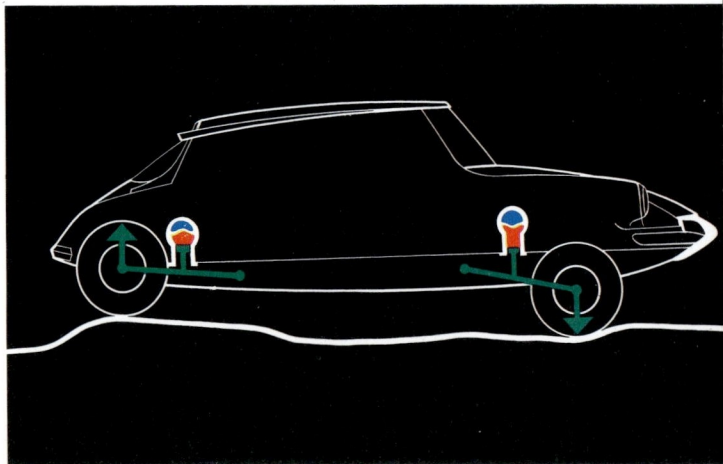

e Änderung der Höhe zwischen
osserie und Boden (Änderung der
stung) setzt den automatischen
enkorrektor in Tätigkeit:
Verbindungsflüssigkeit zwischen
en und Gas wird vermehrt oder
indert die Bodenfreiheit ist
er normal (16 cm).

Mit Hilfe eines einfachen Hebels kann
der Fahrer die Bodenfreiheit nach
Belieben bis auf 28 cm erhöhen, wenn
ihm dies zum Überqueren einer
sehr unebenen Strecke oder einer Furt
notwendig erscheint.

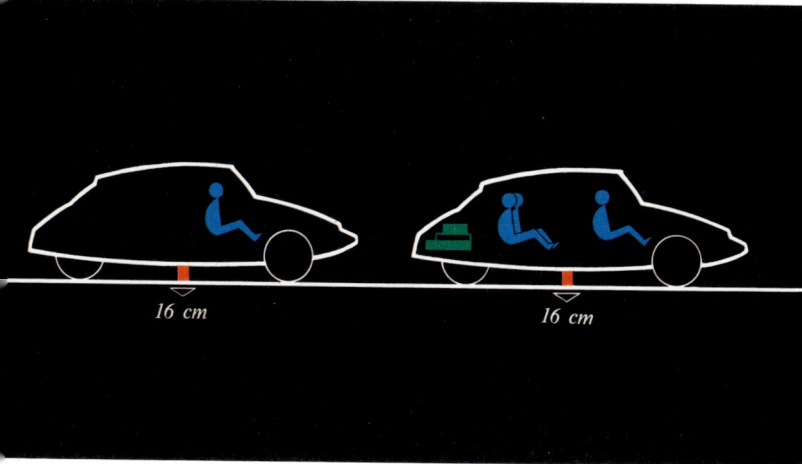

16 cm

16 cm

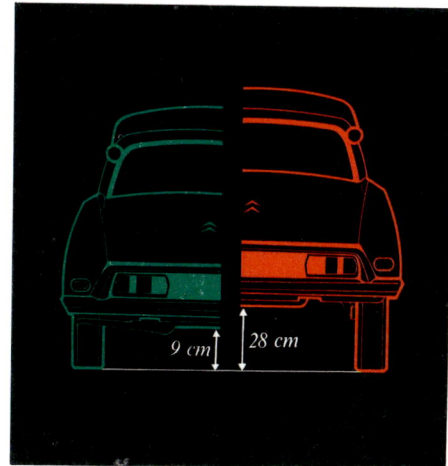

9 cm   28 cm

ds

Die aerodynamische Form des Vorderwagens, die das Rad überdeckenden Kotflügel, das pfeilförmige Vorderteil, die tiefgezogene Motorhaube, die gewölbte Windschutzscheibe, das kurze Heck, sowie das Fehlen unnötiger Erhebungen verringern den Luftwiderstand und verbessern den cw-Wert. Die absolute

Stromlinienform verhindert die Bildung von Luftwirbeln unter der Karosserie und verringert den "Auftrieb", eine Kraft, die bei grossen Geschwindigkeiten dazu neigt, den Wagen vom Boden abzuheben. Die Karosserielinie bietet wenig Angriffsfläche für den Wind.

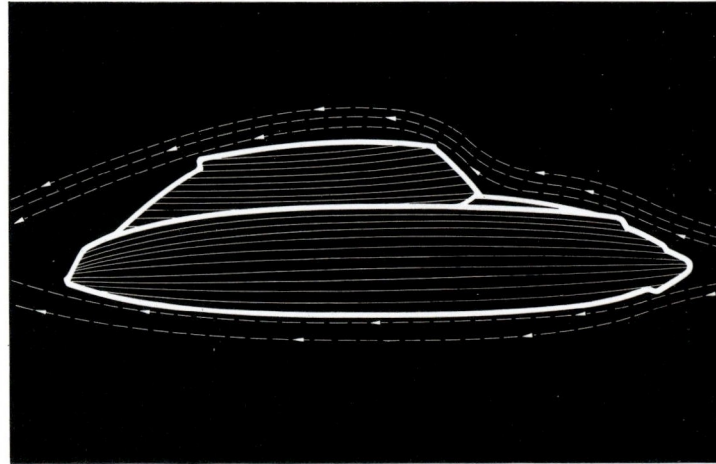

Kühlerverkleidung, keine
lüssigen Lufteintrittsöffnungen,
en Wagen beträchtlich verlang-
n könnten. Zwei Lufteintritte
der Karosserie zum Kühlen der
ibenbremsen. Ein Lufteintritt
zur Motorkühlung: Von einem
en Schacht wird der Luftstrom
Kühler geleitet. Die Frischluft

für den Passagierraum strömt ein
durch je eine kleine Öffnung unter
den Scheinwerfern und wird durch
einen Luftkanal direkt zu den
Lüftungsgittern an beiden Enden
des Armaturenbrettes geleitet.

Die Windschutzscheibe ist stark gewölbt.
Ihr oberer Rand ist gegenüber der unteren
Befestigung erheblich zurückgesetzt und gibt
dadurch vollkommen freie Sicht. Die sehr
schmalen Seitenstreben sind stark nach hinten
versetzt. Ihre Breite wurde dem Gesichtswinkel
entsprechend berechnet. Sie unterbrechen in
keiner Weise den Verlauf der Windschutzscheibe
zu den Türfenstern, die ohne Rahmen und ohne
Ausstellfenster gearbeitet sind.

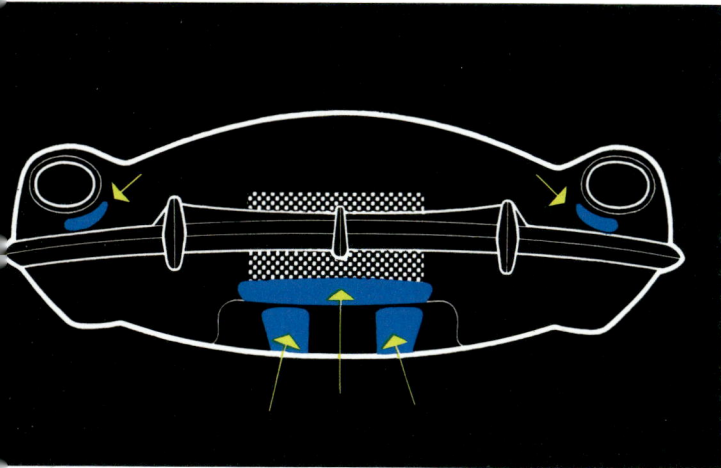

frontantrieb

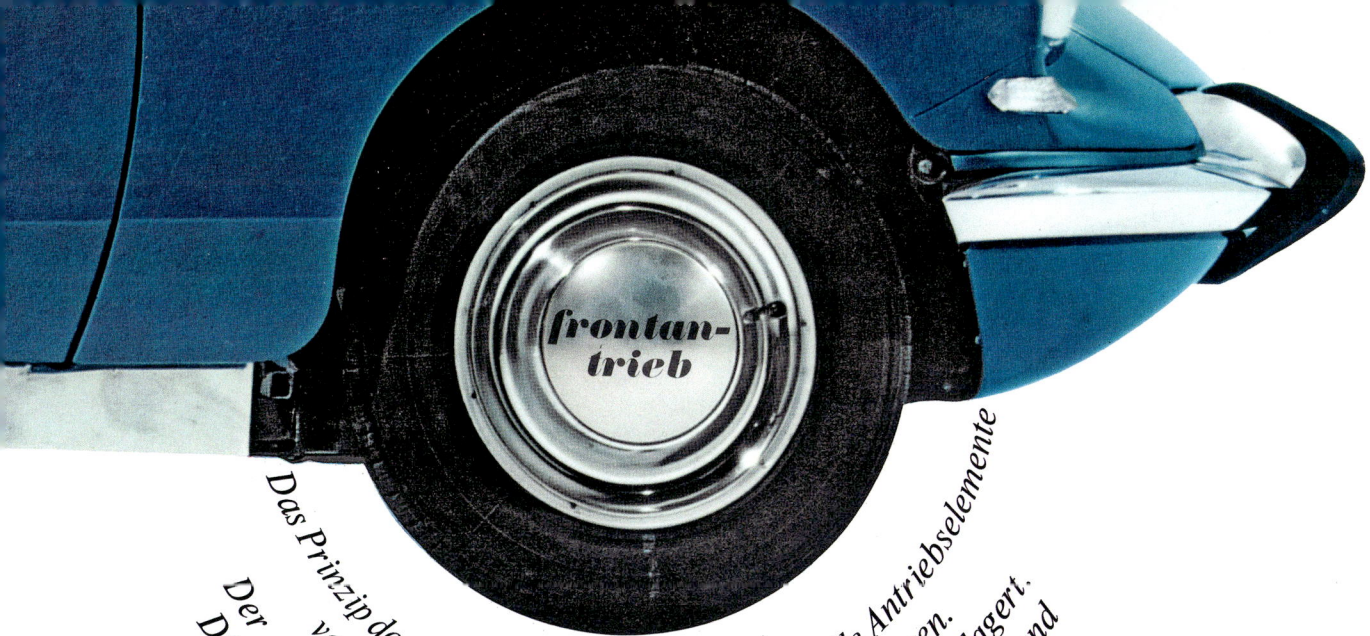

Das Prinzip des Frontantriebs besteht darin, alle Antriebselemente vorn unter der Motorhaube zusammenzufassen. Der Schwerpunkt des Wagens wird dabei nach vorn verlagert. Die Bodenhaftung der Vorderräder, die gleichzeitig Führungs- und Antriebsräder sind, wird entscheidend verbessert. In der Kurve "folgt" die Hinterachse eines Wagens mit Frontantrieb der schwereren Vorderachse und wird durch die Kurve gezogen. Das Fahrzeug "bricht nicht aus". Diese Lösung, zu der sich nach und nach die Konstrukteure der ganzen Welt bekennen, verwirklichte Citroën schon im Jahre 1934. Die technische Ausführung haben Citroën-Konstrukteure seitdem ständig weiter vervollkommnet.

aerodynamik : *Der DS 19 entspricht in seiner äusseren Form und seiner inneren Au*

transportieren. Die Karosserie ist massgerecht um die Fahrgäste geschneidert; von der

DS 19 sieht mehr als andere Autofahrer. Das pfeilförmige Vorderprofil,die fliehenden Linie

auf das Problem des Luftwiderstandes.Die innere Aerodynamik wurde ebenso sorgfältig b

vollendet seiner Bestimmung als Automobil: mehrere Personen schnell, bequem und sicher zu …eiten und bequemen Sitzplätzen aus bietet sich eine freie Sicht nach allen Seiten. Der Fahrer eines …osserie, die vollkommene Windschlüpfigkeit des Wagenkastens: das ist die Antwort von Citroën …tigt nach gründlichem Studium des Lufteintrittes und der Luftzirkulation unter der Motorhaube.

Bei der Hydropneumatik wurden die herkömmlichen Stahlfedern durch vier Gaspolster ersetzt, die

unkomprimierbaren Verbindungsflüssigkeit und einer konstanten komprimierbaren Gasmer
unabhängig aufgehängten Räder reagiert getrennt auf die Unebenheiten des Bodens, während
behält. Einmalig auf der Welt bietet die hydropneumatische Federung eine vollkommene Stra

...ielfaches wirksamer sind als der elastischste Stahl. Es addiert sich dabei die ausgleichende Wirkung einer

Erfolg ist eine sehr flexible, dabei aber völlig stossfreie Federung. Jedes der vier voneinander
...gen dank seines automatischen Höhenkorrektors bei jeder Belastung die gleiche Bodenfreiheit

hydropneumatische federung

Appendix

Appendix

Actors

The Visionaries

André Citroën, company founder. The famous logo with its double arrows references the company's origin, namely as the manufacturer of wedge-shaped step-down gears. In 1919, Citroën entered into automobile manufacturing and in 1934 launched the famous Traction Avant, the first industrially produced car with front wheel drive. Unfortunately, that development completely overwhelmed the company's finances. In 1935, André Citroën died a financially ruined man.

The Michelin Brothers, primary investors, took over the factory in 1934 and, together with the new director Pierre Boulanger (until his fatal car accident in 1950) and Robert Puiseux (until 1958), continued progressive and innovative company politics. Together with the best engineers and designers, they would write automobile history.

In 1968, **Citroën** and **Peugeot** merge and, in the mid-1970s, also take over **Maserati**.

The Designers

Flaminio Bertoni, designer, sculptor and architect, born 1903 near Varese, died 1964 in Paris. He had come to Paris the first time in 1923 by invitation of French engineers and moved there permanently in 1931.

As a designer, he completed and modeled the famous Citroën "Traction Avant" in a single night, as he related in his autobiography. Flaminio Bertoni would design three more path-breaking cars for Citroën, namely the 2CV (1948), then the famous Citroën DS (1955) and finally, the Ami 6 (1961) with is characteristic angled rear window.

As a sculptor, Flaminio Bertoni exhibited his work on various occasions, for example in 1936 in Paris with other Italian artists, among them de Chirico. In the 1950s, he made a number of smaller sculptures which recall certain works by Brâncuşi.

As an architect, he completed several renovations in Paris in the 1950s, and in 1952 he designed a house in Boulogne, among other projects.

Henry Chapron, designer and car body builder, built the DS convertible for Citroën based upon Bertoni's design and produced special car bodies in variations on Citroën's models, for example the limousines for various presidents.

Robert Opron, designer, became Flaminio Bertoni's successor at Citroën. They worked collaboratively on modifying the front portion of the DS with the famous moveable headlights beginning in 1968. Robert Opron later designed the GS (1970) and the CX (1974), which was actually the next in line after the DS. Through the merger between Citroën and Maserati, the famous SM (1970) was produced, also designed by Opron.

Studio Bertone in Turin introduces hard-edged form and designs the BX (1982) as successor to the GS, as well as the XM (1989) as successor to the CX.

The Engineers

André Lefèbvre, engineer/constructeur, moved from the famous aircraft manufacturer Voisin in 1929 first to Renault and then two years later to Citroën. Trained as an airplane engineer, he totally reconceived the relationships between chassis and car body on the one hand, and aerodynamics and the dynamics of driving on the other. As constructeur for Citroën, he developed the Traction Avant, the 2CV and finally, the DS.

Paul Magés, autodidact, also known as "The Professor," was hired in 1936 at the age of seventeen by Pierre Boulanger. Through persistent experimentation, he developed the hydropneumatic suspension. Hydropneumatic suspension – the car rests to a certain extent on a pillow of air (nitrogen) and the transmission works via an oil-based system – is integrated into the Traction Avant 15 Six H in early 1954 and facilitates consistent height while driving as well as under asymmetrical loading conditions. In the DS, front and rear axels are on hydropneumatic suspension and are also connected to the drive train and transmission.

Walter Becchia, the "pope of motors," arrived at Citroën in 1946, built the flat box motor for the 2CV and developed the Traction Avant motor further into the form integrated into the DS.

Together with the engineers from Maserati, he later built the motor for the Citroën Maserati / SM in aluminum. The motors were so "overbred" that they ignite on occasion – the cars often have telltale smoke residue.

Chronology

1938 / 1947	Beginning	Studies for the VGD Voiture de grande diffusion, internal nickname "Hippopotame," the hippopotamus
1955 / 1956	DS 19	"Déesse" (Goddess). Motor 191 cm³, 78×100 mm, same motor as the Traction Avant with new hood
1957	ID 19	"Idée" (Idea)
1959		**ID Confort** and ID Break
1959		DS, ID and ID Break get a new tail design (the elongated taillights are replaced by point-shaped ones, the rear fenders are adjusted)
1961	DS and ID	Convertible produced by Chapron
1963		New radiator design (rubber bumper)
1965	DS Pallas	Additional decorative molding, more luxurious interior, etc.
1966	DS / ID 21	Compared to the DS 19, a more powerful motor (2175 cm³, 90×85.5 mm)
1967		New green oil replaces red oil in the hydraulic system
1967 / 1968		DS/ID new design of the front radiator face
1969	DS / ID 20	Improved DS/ID 19 series, same motor
1971		Convertible production terminated
1973		**DS 23** Compared to DS 21, more powerful motor (2347 cm³, 93.5×85.5 mm)
1975		Production terminated

Bibliography "Citroën DS"

Context

- "The New Citroën," in Roland Barthes, *The Eiffel Tower and Other Mythologies,* translated by Richard Howard, Hill and Wang, New York 1979
- Otl Aicher, *Kritik am Auto. Schwierige Verteidigung des Autos gegen seine Anbeter,* Callwey, München 1984
- Alison Smithson, *AS in DS. An Eye on the Road,* Christian Sumi (ed.), reprint, Lars Müller Publishers, Baden 2001

Flaminio Bertoni

- *Flaminio Bertoni,* Macchione Editore, Varese 1997
- Leonardo Bertoni, *Flaminio Bertoni – La vita, il Genio e le Opere,* Macchione Editore, Varese 2002

The Car

- Jon Presnell, *Citroën DS. The Complete Story,* Crowood Auto Classic, The Bath Press, Ramsbury 1999
- Jan de Lange, John Reynolds, *Citroën DS, alle DS und ID Modelle von 1955–75,* Heel AG, Schindellegi, Schweiz 1997, German edition
- Shirine Guy, Antoine Demetz (ed.), *DS toujours d'avant-garde / DS – Always Avant-Garde,* Citroën communication, Gutenberg Networks, France 2015
- Roger Brioult, *Citroën. L'histoire et les secrets de son bureau d'études,* Collection "Histoires d'Autos", n. 5, ediFree "La vie de l'auto," Fontainebleau 1987
- Fabien Sabatès, Didier Lainé, *Citroën DS 1955–1975,* Collection Auto Archives no. 13 de 1984
- Hans Otto Meyer-Spelbrink, *Das ungewöhnlichste Serienautomobil aller Zeiten,* Podszun Verlag, Brilon 2003
- Jacques Borgé, Nicolas Viasnoff, *L'Album de la DS,* EPA, Nancy 1986

Streamline Form

- Ralf J.F. Lieselbach, *Stromlinienautos in Europa und den USA. Aerodynamkik im PKW-Bau 1900–1945,* Kohlhammer Edition Auto und Verkehr, Stuttgart 1982
- Raymond Loewy, *Hässlichkeit verkauft sich schlecht,* Econ Verlag, Düsseldorf 1992, German edition
- Raymond Loewy, *Industrial Design,* Faber and Faber, London, Boston 1979
- *Studebaker 1946–1958,* Howard L. Applegate (ed.), Iconografix Photo Archive Series, Minnesota 1995
- Claude Lichtenstein, Franz Engeler (ed.), *Stromlinienform,* Museum für Gestaltung / Verlag Lars Müller, Zürich 1992

Gestalt and Form

- Rudolf Arnheim, *Art and Visual Perception. A Psychology of the Creative Eye,* University of California Press, Berkeley 1954. Second Edition, University of California Press, Berkeley 1974

– Martin Steinmann, *Forme forte, Schriften / Ecrits 1972–2002*, Jacques Lucan, Bruno Marchand (eds.), Birkhäuser Verlag, Basel 2003

Max Bill
– Max Bill, "Schönheit aus Funktion und als Funktion," in: Werk 8 / 1949, pp. 272–274
– *Max Bill: Form. Eine Bilanz über die Formentwicklung um die Mitte des XX. Jahrhunderts*, Karl Werner, Basel 1952
– Max Bill, "(Form, Funktion, Schönheit) = (Gestalt)" in: *Max Bill,* exhibition catalog, Ulm 1956
– *minimal tradition. Max Bill und die "einfache" Architektur 1942–1996*, XIX Triennale di Milano, Bundesamt für Kultur (ed.), Stanislaus von Moos (commissioner), Verlag Lars Müller, Baden 1996
– Lars Müller (ed.), *Max Bill's View of Things. Die gute Form: An Exhibition 1949*, Lars Müller Publishers, Zürich 2015, p. 143 ff.

Michel Zumbrunn (selection)
– Michel Zumbrunn, *Auto legends. Classics of Style and Design*, Text Robert Cumberford, Merrell, London New York 2004
– Michel Zumbrunn, *Stromlinie*, Motorbuch, Stuttgart 2012

– *The Ferrari Book. Passion for Design*, Michael Köckritz (ed.), Jürgen Lewandowski (author), Michel Zumbrunn (photographer), teNeues Media, Kempen 2017

Bibliography "Fragments of the Modern"

Flaminio Bertoni

- Christian Sumi, *The Goddess – La Déesse. Investigations on the legendary Citroën DS,* Lars Müller Publishers, Zürich 2020
- Christian Sumi (ed.), Alison Smithson: *AS in DS. An eye on the road,* reprint 2001, Lars Müller Publishers, Baden 2001

Le Corbusier

- "The immeuble Clarté," in: *In The Footsteps of le Corbusier,* Rizzoli, New York 1991, pp. 177–187
- Christian Sumi: *Immeuble Clarté Genf 1932 von Le Corbusier und Pierre Jeanneret. Maison à sec · Immeuble-villa · plan libre,* gta / Ammann Verlag, Zürich 1989
- "Clarté (immeuble) und Wanner (Edmond)," in: *Le Corbusier, une encyclopédie, monographie,* Centre Georges Pompidou, Paris 1987, pp. 100, 477–478
- "L'immeuble Clarté et la conception de la maison à sec," in: *Le Corbusier à Genève 1922–32,* Verlag Payot, Lausanne 1987, pp. 93–111
- "Vom Mehrfamilienhaus konzipiert als Villas Superposées zum Mehrfamilienhaus als kollektives Wohnhaus," in: *Le Corbusier – La Ricerca paziente,* FAS Ticino Lugano 1980, pp. 63–68
- Christian Sumi: "Il Progetto Wanner" in: *Rassegna,* 3/1980, pp. 39–46

Caccia Dominioni

- Christian Sumi, Luigi Caccia Dominioni "Schleier und Kontext," in: *werk, bauen + wohnen,* 12-2013, pp. 18–19

Mies van der Rohe

- Christian Sumi, "Fragments of the modern. Mies van der Rohe," in: *Giornale IUAV:164,* Venezia 2019, pp. 6, 7
- Marianne Burkhalter, Christian Sumi, "Mies van der Rohe, Three Plaza projects 1964–1969," in: *2G – 35, 2005,* pp. 142, 143

Giulio Minoletti

- Cristina Loi, Christian Sumi, Annalisa Viati (eds.), *Giulio Minoletti 1910–1981,* Archivio del Moderno (AdM), Mendrisio Academy Press / SilvanaEditoriale, Milano 2017
- Giulio Minoletti at Archivio del Moderno, "Vom Archiv zum Projekt," in *domus 006* (German edition), März/April 2014, pp. 46–47
- Christian Sumi, Annalise Viati Navone (author), *Giulio Minoletti – architetto, urbanista e designer* Mendrisio Academy Press, Archivio del Moderno, 2014
- Christian Sumi, "Kind of Blue. Arcadia Garden von Giulio Minoletti in Mailand 1959," in: *L'opera sovrana, studi*

sull'architettura del XX secolo dedicati a Bruno Reichlin, Archivio del Moderno (AdM), Mendrisio Academy Press / SilvanaEditoriale, Milano 2014, pp. 404–415
- Giulio Minoletti at Archivio del Moderno, "Dall'archivio al progetto," in: *domus 973*, October 2013, pp. 23–25

Jean Prouvé
- Christian Sumi: "Die Verwendung von Holz und das Haus BCC," in: *Jean Prouvé*, Verlag Vitra Design Museum, Weil a.R., 2005, pp. 194–199

Otto Rudolf Salvisberg
- Christian Sumi: "Spitalbauten," in: *O.R. Salvisberg – die andere Moderne*. gta Verlag, Zürich 1985, pp. 196–205
- Christian Sumi (together with Ernst Strebel): "Detailzeichnungen," in: *O.R. Salvisberg – die andere Moderne*: gta Verlag, Zürich 1985, pp. 206–219

Rudolf Steiger (HMS) and Flora Crawford
- Marianne Burkhalter, Christian Sumi (eds.), *Haus Steiger 1959*, gta Verlag ETH Zürich 2021

Gottfried Semper
- Christian Sumi, "Öffentlichkeit im Raum: Gottfried Semper – 3 Forumsprojekte in 6 Modellen," in: *werk, bauen + wohnen*, 10/2003, pp. 33–39

- Christian Sumi, "Pompei in Zürich," in: *archithese*, 5/2003, pp. 28–31

Konrad Wachsmann
- Christian Sumi, "Konrad Wachsmann – Pioneer of Industrial Building," in: *Shukhov. The Formula of Architecture*, Schusev State Museum of Architecture, Moscow 2019, pp. 344–357
- Marianne Burkhalter, Christian Sumi (eds.), *Konrad Wachsmann and the Grapevine Structure*, Park Books, Zürich 2018
- Christian Sumi, "Holzhausbau heute," in: *Konrad Wachsmann. Holzhausbau*, Birkhäuser Verlag, Basel 1995 (reprint from 1930), pp. 19–27/"Building the Wooden House Today" in: *Konrad Wachsmann, Building the Wooden House*, Birkhäuser Verlag, Basel 1995, English edition, pp. 19–27

Marco Zanuso
- Christian Sumi, Marco Zanuso, "Tectonic and Gestalt," in: *Marco Zanuso. Architettura e Design*, Luciano Crespi, Letizia Tedeschi, Annalisa Viati Navone, a cura di Officina libraria, Milano 2020
- Christian Sumi, "Forma e percezione: i televisori Doney, Algol e Black per Brionvega," in: *MZ Progetto integrato, Marco Zanuso, design, technica e industria*, Mendrisio Academy Press, Archivio del Moderno, 2013, pp. 54–65

Car Graveyard

Photos: Christian Sumi

Car Graveyard Photos: Christian Sumi

The car graveyard is in Boucoiran, France, between Nîmes and Alès. Like a herd of stampeding buffalos, the cars are in tight formation on a meadow overgrown with tall grass. The grass conceals the tires, making the car bodies into floating objects without wheels, which would have been Bertoni's ideal condition. Even in this new context, the form "asserts" itself through its formal qualities.

The car "conquers" nature and nature "conquers" the car: the car as part of nature or nature as part of the car. The glass cladding on the headlights and the passenger compartment become greenhouses as mold covers the bodies.

Thanks

I would like to thank Michel Zumbrunn and Heinz Unger for the comprehensive photo material of the car, they form the basis of the investigation. Furthermore my thanks go to the editors of the core literature *Stromlinienform (Streamlined Form)*, Claude Lichtenstein and Franz Engler, for reading the manuscript and contributing complementary references; to Martin Steinmann for the thematic cluster around "Gestalt and perception." I would also like to thank Daniel Kunz (Häfliger & Kunz AG Citroën-Restaurationen) for the technical clarification and explanations; burkhalter sumi architekten for supporting the graphic work; Lars Müller Publishers; Karin Schiesser and Paul Märki for their graphic design; and Lynnette Widder for the translation.

About the author

Christian Sumi graduated from the Zurich Polytechnic ETH, where he worked for the Institute of History and Theory of Architecture (gta). He has taught at Harvard University GSD, the Lausanne Polytechnic EPFL and the University of Strathclyde in Glasgow. From 2008 to 2016 he had a full professorship at the Accademia di Architettura Mendrisio AAM, together with Marianne Burkhalter. Since 1984, Sumi and Burkhalter have been running an architectural practice in Zurich. The firm acquired an international reputation with, among other things, transformations of existing structures, new housing typologies, hotels and innovative wooden constructions and their distinctive use of polychrome colors. The office also made a number of exhibition installations: *Gottfried Semper* and *Robots* for the Museum für Gestaltung Zürich (2004 and 2009); *The unexpected view* and *Konrad Wachsmann – the Grapevine Structure* for La Biennale di Venezia (2014 and 2018). Sumi is the editor of the reprint of Alison and Peter Smithson's publication *AS in DS*, published by Lars Müller Publishers (2001).

The Goddess – La Déesse

Investigations on the Legendary Citroën DS

Author: Christian Sumi
Photography: Michel Zumbrunn, Heinz Unger
Translation: Lynnette Widder
Proofreading: Stephanie Shellabear
Coordination: Maya Rüegg
Design: Karin Schiesser
Assistance: Paul Märki
Production: Martina Mullis
Lithography: prints professional, Berlin, Germany
Printing and binding: Printer Trento, Italy
Paper: Magno Volume, 130 gsm

Lars Müller Publishers is supported by the
Swiss Federal Office of Culture with a structural
contribution for the years 2016–2020.

Lars Müller Publishers
Zürich, Switzerland
www.lars-mueller-publishers.com

ISBN 978-3-03778-626-0

Distributed in North America by ARTBOOK | D.A.P.
www.artbook.com

Printed in Italy